# NOVEL FOOD
# FERMENTATION TECHNOLOGIES

# NOVEL FOOD FERMENTATION TECHNOLOGIES

Evangelia Varella
Gilberto Artioli

Kruger Brentt
Publishers
2023

Kruger Brentt Publishers UK. LTD.
Company Number 9728962

Regd. Office: 68 St Margarets Road, Edgware, Middlesex HA8 9UU



ISBN: 9781787150805



For information on all our publications visit our website at http://krugerbrentt.com/

# PREFACE

Fermentation is a natural process through which microorganisms like yeast and bacteria convert carbs — such as starch and sugar — into alcohol or acids. The alcohol or acids act as a natural preservative and give fermented foods a distinct zest and tartness. Fermentation also promotes the growth of beneficial bacteria, known as probiotics. Fermentation is a biochemical reaction that extracts energy from carbohydrates without using oxygen. Under anaerobic conditions, microbes break down sugar into alcohol, acid, and/or gas to produce adenosine triphosphate to stay alive, but the microbes also produce ethanol, carbon dioxide, and lactic acid, which increase product shelf life and impart desirable sensory properties.Food fermentation is a food processing technology that utilizes the growth and metabolic activity of microorganisms for the stabilization and transformation of food materials. Fermentation was primarily developed for the stabilization of perishable agricultural produce. Notwithstanding, the technology has evolved beyond food preservation into a tool for creating desirable organoleptic, nutritional, and functional attributes in food products.

Fermented foods contain healthy live bacteria known as probiotics and are foods that have gone through a process during which this bacteria converts the starches and sugars in that food into lactic acid and acetic acid. Fermented foods have high nutritional values (vitaminK2, trace minerals, B-vitamins and probiotics), are easy to prepare and are economical. Fermentation is an old food preservation method and was used by the Romans who consumed sauerkraut, Ancient Indians who enjoyed " Lassi" a pre-dinner yogurt drink, Bulgarians who are known for their high consumption of fermented milk and Kefir, Turks who are famous for their Ariana drink alongside their meals, Asians who pickle cabbage, turnips, eggplant, cucumbers, onions, squash, and carrots and Middle Easterners who use yoghurt to accompany almost every meal. Future trends in fermentation will include a greater focus on natural fermentation and increased growth in the craft beer, kombucha, vegetable fermentation, yogurt, and kefir markets. There will also be increased growth in fermented food markets as researchers learn more about the relationship between the composition of different fermented foods, the human gut microbiome, and human health.

The present book contains eleven covering all related disciplines. These chapters include introduction, fermentation technology, bio-fermentation technology, fermenters and bioreactors, food fermentation and micro-organisms, industrial food fermentation, beer and wine fermentation, fermentation of dairy products, fermented cereals and emulsions & gels fermentation and fermented foods processing-flavor, preservation, and health. It provides analytical details and a comprehensive overview of innovations in food fermentation technologies. This book also the basics of fermentations, types of fermentations described in broad perspective. The book also explains significance of fermented food in nutrition and overview of the different starter cultures and health benefits of fermented food products. This book would be an invaluable asset for the scientists, students and teachers working in the field of biotechnology, agriculture, microbiology, agriculture engineering, food technology and community science.

We express indebtedness to authors of books, manuals, periodicals, magazines, newsletters, journals and websites etc. that helped in the preparation of this book. It has not been possible to obtain permission for reproduction of some of the materials for which authors and publishers offer their sincere apologies. Healthy criticism and suggestions for further improvement of the book are solicited.

**Evangelia Varella**

**Gilberto Artioli**

# CONTENTS

# 1

# INTRODUCTION

Food fermentation is the process of creating food or changing the properties of food using microbes. Many cultures started fermenting foods to preserve them. For example, fermenting vegetables allowed people living in places with harsh winters to eat them year-round. Cheese is another fermented food that lasts much longer than its previous form, milk. For some foods, like olives, fermentation makes an inedible or bad-tasting food edible or more palatable.

Fermentation is a metabolic process that produces chemical changes in organic substrates through the action of enzymes. In biochemistry, it is narrowly defined as the extraction of energy from carbohydrates in the absence of oxygen. In food production, it may more broadly refer to any process in which the activity of microorganisms brings about a desirable change to a foodstuff or beverage. The science of fermentation is known as zymology.

In microorganisms, fermentation is the primary means of producing adenosine triphosphate (ATP) by the degradation of organic nutrients anaerobically. Humans have used fermentation to produce foodstuffs and beverages since the Neolithic age. For example, fermentation is used for preservation in a process that produces lactic acid found in such sour foods as pickled cucumbers, kombucha, kimchi, and yogurt, as well as for producing alcoholic beverages such as wine and beer. Industrial fermentation is the process of using microbes for the large-scale production of chemicals, biofuels, enzymes, proteins and pharmaceuticals. Fermentation also occurs within the gastrointestinal tracts of all animals, including humans.

Along with aerobic respiration, fermentation is a method to extract energy from molecules. This method is the only one common to all bacteria and eukaryotes. It is therefore considered the oldest metabolic pathway, suitable for primeval environments – before plant life on Earth, that is, before oxygen in the atmosphere. :

Yeast, a form of fungus, occurs in almost any environment capable of supporting microbes, from the skins of fruits to the guts of insects and mammals to the deep ocean. Yeasts convert sugar-rich molecules to produce ethanol and carbon dioxide.

Basic mechanisms for fermentation remain present in all cells of higher organisms. Mammalian muscle carries out fermentation during periods of intense exercise where oxygen supply becomes limited, resulting in the creation of lactic acid.: In invertebrates, fermentation also produces succinate and alanine.

Fermentative bacteria play an essential role in the production of methane in habitats ranging from the rumens of cattle to sewage digesters and freshwater sediments. They produce hydrogen, carbon dioxide, formate and acetate and carboxylic acids. Then consortia of microbes convert the carbon dioxide and acetate to methane. Acetogenic bacteria oxidize the acids, obtaining more acetate and either hydrogen or formate. Finally, methanogens (in the domain Archea) convert acetate to methane.

Fermented foods and beverages are becoming increasingly popular. You can now find kombucha on tap in bars, sourdough bread in cafes and more varieties of yoghurt, sauerkraut and kimchi in your local supermarket.

Fermented foods have been part of the human diet for centuries, and were initially produced as a way to preserve foods, improve flavour and eliminate food toxins. Today, more people are turning to these foods for their potential health benefits.

## 1.1 FERMENTATION IN FOOD PROCESSING

In food processing, fermentation is the conversion of carbohydrates to alcohol or organic acids using microorganisms—yeasts or bacteria—under anaerobic (oxygen-free) conditions. Fermentation usually implies that the action of microorganisms is desired. The science of fermentation is known as zymology or zymurgy.

The term, "fermentation" sometimes refers specifically to the chemical conversion of sugars into ethanol, producing alcoholic drinks such as wine, beer, and cider. However, similar processes take place in the leavening of bread ($CO_2$ produced by yeast activity), and in the preservation of sour foods with the production of lactic acid, such as in sauerkraut and yogurt.

Other widely consumed fermented foods include vinegar, olives, and cheese. More localised foods prepared by fermentation may also be based on beans, grain, vegetables, fruit, honey, dairy products, and fish.

### 1.1.1 History and Prehistory

The deliberate fermentation of foods by man predates written history and is possibly the oldest method of preserving perishable foods. Evidence suggests that fermented foods were consumed 7,000 years ago in Babylon. Scientist speculate that our ancestors possibly discovered fermentation by accident and continued to use the process out of preference or necessity. Preserving by fermentation not only made foods available for future use, but more digestible and flavorful. The nutritional value produced by fermenting is another benefit of fermenting.

The study of fermentation is called zymology. You may, or may not, know the list of foods in our present culture that have been fermented for centuries, at some stage of production, to be made more edible and accessible to the human population.

Natural fermentation precedes human history. Since ancient times, humans have exploited the fermentation process. The earliest archaeological evidence of fermentation is 13,000-year-old residues of a beer, with the consistency of gruel, found in a cave near Haifa in Israel. Another early alcoholic drink, made from fruit, rice, and honey, dates from 7000 to 6600 BC, in the Neolithic Chinese village of Jiahu, and winemaking dates from ca. 6000 BC, in Georgia, in the Caucasus area. Seven-thousand-year-old jars containing the remains of wine, now on display at the University of Pennsylvania, were excavated in the Zagros Mountains in Iran. There is strong evidence that people were fermenting alcoholic drinks in Babylon ca. 3000 BC, ancient Egypt ca. 3150 BC, pre-Hispanic Mexico ca. 2000 BC, and Sudan ca. 1500 BC.

The French chemist Louis Pasteur founded zymology, when in 1856 he connected yeast to fermentation. When studying the fermentation of sugar to alcohol by yeast, Pasteur concluded that the fermentation was catalyzed by a vital force, called "ferments", within the yeast cells. The "ferments" were thought to function only within living organisms. "Alcoholic fermentation is an act correlated with the life and organization of the yeast cells, not with the death or putrefaction of the cells", he wrote.

Nevertheless, it was known that yeast extracts can ferment sugar even in the absence of living yeast cells. While studying this process in 1897, the German chemist and zymologist Eduard Buchner of Humboldt University of Berlin, Germany, found that sugar was fermented even when there were no living yeast cells in the mixture, by an enzyme complex secreted by yeast that he termed zymase. In 1907 he received the Nobel Prize in Chemistry for his research and discovery of "cell-free fermentation". One year earlier, in 1906, ethanol fermentation studies led to the early discovery of NAD+.

### 1.1.2 Antiquity

Charred crumbs of a flatbread made by Natufian hunter-gatherers from wild wheat, wild barley and plant roots between 14,600 and 11,600 years ago have been found at the archaeological site of Shubayqa 1 in the Black Desert in Jordan, predating the earliest known making of bread from cultivated wheat by thousands of years. Grinding stones dated at 30,000 years old, possibly used for grinding grains and seeds into flour, have in recent years been unearthed in Australia and Europe, but there is no definitive evidence that these tools or their products were used for making breads.

Bread is otherwise strongly associated with agriculture. Wheat was domesticated in the Fertile Crescent. Bread is found in Neolithic sites in Turkey and Europe from around 9,100 years ago

There is extensive evidence of breadmaking in Ancient Egypt in the form of artistic depictions, remains of structures and items used in bread making, and remains of the dough and bread itself.

The most common source of leavening in antiquity was to retain a piece of dough from the previous day to utilize as a form of sourdough starter. Pliny the Elder reported that the Gauls and Iberians used the foam skimmed from beer to produce "a lighter kind of bread than other peoples." Parts of the ancient world that drank wine instead of beer used a paste composed of grape must and flour that was allowed to begin fermenting, or wheat bran steeped in wine, as a source for yeast. The idea of a free-standing oven that could be pre-heated, with a door for access, appears to have been Greek.

Even in antiquity there was a wide variety of breads. In ancient times the Greek bread was barley bread: Solon declared that wheaten bread might only be baked for feast days. By the 5th century BC bread could be purchased in Athens from a baker's shop, and in Rome, Greek bakers appeared in the 2nd century BC, as Hellenized Asia Minor was added to Roman dominion as the province of Asia; the foreign bakers of bread were permitted to form a collegium. In the Deipnosophistae, the author Athenaeus describes some of the bread, cakes, and pastries available in the Classical world. Among the breads mentioned are griddle cakes, honey-and-oil bread, mushroom-shaped loaves covered in poppy seeds, and the military specialty of rolls baked on a spit. The type and quality of flours used to produce bread could also vary, as noted by Diphilus when he declared "bread made of wheat, as compared with that made of barley, is more nourishing, more digestible, and in every way superior. In order of merit, the bread made from refined flour comes first, after that bread from ordinary wheat, and then the unbolted, made of flour that has not been sifted." The essentiality of bread in the diet was reflected in the name for the rest of the meal: ópson, "condiment", i.e. bread's accompaniment, whatever it might be.

### 1.1.3 Middle Ages

In medieval Europe, bread served not only as a staple food but also as part of the table service. In the standard table setting of the day the trencher, a piece of stale bread roughly 6 inches by 4 inches (15 cm by 10 cm), was served as an absorbent plate. When food was scarce, an all-too-common occurrence in medieval Europe, the trencher when served would typically be eaten with or at the completion of a meal. In times of relative abundance, trenchers could be given to the poor or fed to the dogs. It was not until the 15th century that trenchers made of wood started to replace the bread variety. In Britain the price, weight and quality of bread and beer was regulated by the Assize of Bread and

Ale from the 13th century. Assizes were abandoned in Glasgow in 1801, in London in 1815 and in the rest of Britain in 1836.

## 1.2 TO THE 19TH CENTURY

The Corn Laws inflated the price of bread. The Anti-Corn Law League demanded cheap bread. After repeal of the Corn Laws in 1846 cereal duties were substantially reduced, and abolished in 1869.

From the late 18th century to the end of the 19th century, bread sold in England was often adulterated with hazardous materials, including chalk, sawdust, alum, plaster, clay and ammonium carbonate. Frederick Accum was the first to raise alarm to the food adulteration in 1820.

This gradually came to an end with government action, such as the 1860 and 1899 Food Adulteration Acts in Britain. America had a more difficult time ending these processes of adulteration, however, as various states had varying policies regarding bread making. In the mid-19th century, Britain imported much of its bread wheat from the United States.

### 1.2.1 Industrialization

Bread-baking was industrialised at the start of the 20th century. Otto Frederick Rohwedder developed a prototype bread-slicing machine in 1912, and a practical machine that both sliced and wrapped bread in 1928.

It was discovered early on that while a bran- and wheatgerm-discarding milling process can help improve white flour's shelf life, it does remove nutrients like some dietary fiber, iron, B vitamins, micronutrients and essential fatty acids. The US government has mandated since 1941 fortification of white flour-based foods with some of the nutrients lost in milling, like thiamin, riboflavin, niacin, and iron. This mandate came about in response to the vast nutrient deficiencies seen in US military recruits at the start of World War II.

A major change in the United Kingdom was the development in 1961 of the Chorleywood bread process. This used the intense mechanical working of dough, and control of gases touching dough, to dramatically reduce the fermentation period and the time taken to produce a loaf at the expense of taste and nutrition.

For generations, white bread was the preferred bread of the rich while the poor ate dark (whole-grain) bread. However, in most western societies, the connotations reversed in the late 20th century, with whole-grain bread becoming preferred as having superior nutritional value while Chorleywood bread became associated with lower-class ignorance of nutrition.

More recently, and especially in smaller retail bakeries, chemical additives are used that both speed up mixing time and reduce necessary fermentation time, so that a batch of bread may be mixed, made up, risen, and baked in fewer than three hours. Dough that does not require fermentation because of chemical additives is called "quick bread" by commercial bakers. Common additives include reducing agents such as L-cysteine or sodium metabisulfite, and oxidants such as potassium bromate or ascorbic acid; this last ingredient is added to wholemeal bread to increase the softness of the loaf. Calcium was added to flour in the UK to prevent rickets which had been detected as common in women who joined the World War II effort. Since 1986, domestic breadmakers that automate the process of making bread have become popular in the home.

## 1.3 DEVELEPMENT OF BIOCHEMISTRY IN FOOD PROCESSING AREAS

The history of biochemistry can be said to have started with the ancient Greeks who were interested in the composition and processes of life, although biochemistry as a specific scientific discipline has its beginning around the early 19th century. Some argued that the beginning of biochemistry may have been the discovery of the first enzyme, diastase (today called amylase), in 1833 by Anselme Payen, while others considered Eduard Buchner's first demonstration of a complex biochemical process alcoholic fermentation in cell-free extracts to be the birth of biochemistry. Some might also point to the influential work of Justus von Liebig from 1842, Animal chemistry, or, Organic chemistry in its applications to physiology and pathology, which presented a chemical theory of metabolism, or even earlier to the 18th century studies on fermentation and respiration by Antoine Lavoisier.

The term "biochemistry" itself is derived from the combining form bio-, meaning "life", and chemistry. The word is first recorded in English in 1848, while in 1877, Felix Hoppe-Seyler used the term Biochemie in German in the foreword to the first issue of Zeitschrift für Physiologische Chemie as a synonym for physiological chemistry and argued for the setting up of institutes dedicate to its studies. Nevertheless, several sources cite German chemist Carl Neuberg as having coined the term for the new discipline in 1903, and some credit it to Franz Hofmeister.

The subject of study in biochemistry is the chemical processes in living organisms, and its history involves the discovery and understanding of the complex components of life and the elucidation of pathways of biochemical processes. Much of biochemistry deals with the structures and functions of cellular components such as proteins, carbohydrates, lipids, nucleic acids and other biomolecules; their metabolic pathways and flow of chemical energy through metabolism; how biological molecules give rise to the processes that occur within living cells; it also focuses on the biochemical processes involved in the control of information flow through biochemical signalling, and how they relate to the functioning of whole organisms. Over the last 40 years the field has had success in explaining living processes such that now almost all areas of the life sciences from botany to medicine are engaged in biochemical research.

Among the vast number of different biomolecules, many are complex and large molecules (called polymers), which are composed of similar repeating subunits (called monomers). Each class of polymeric biomolecule has a different set of subunit types. For example, a protein is a polymer whose subunits are selected from a set of twenty or more amino acids, carbohydrates are formed from sugars known as monosaccharides, oligosaccharides, and polysaccharides, lipids are formed from fatty acids and glycerols, and nucleic acids are formed from nucleotides. Biochemistry studies the chemical properties of important biological molecules, like proteins, and in particular the chemistry of enzyme-catalyzed reactions. The biochemistry of cell metabolism and the endocrine system has been extensively described. Other areas of biochemistry include the genetic code (DNA, RNA), protein synthesis, cell membrane transport, and signal transduction.

### 1.3.1 Proto-Biochemistry

In a sense, the study of biochemistry can be considered to have started in ancient times, for example when biology first began to interest society—as the ancient Chinese developed a system of medicine based on yin and yang, and also the five phases, which both resulted from alchemical and biological interests. Its beginning in the ancient Indian culture was linked to an interest in medicine, as they developed the concept of three humors that were similar to the Greek's four humours. They also delved into the interest of bodies being composed of tissues. The ancient Greeks conception of "biochemistry" was linked with their ideas on matter and disease, where good health was thought to come from a balance of the four elements and four humors in the human body. As in the majority of early sciences, the Islamic world contributed significantly to early biological advancements as well as alchemical advancements; especially with the introduction of clinical trials and clinical pharmacology presented in Avicenna's The Canon of Medicine. On the side of chemistry, early advancements were heavily attributed to exploration of alchemical interests but also included: metallurgy, the scientific method, and early theories of atomism. In more recent times, the study of chemistry was marked by milestones such as the development of Mendeleev's periodic table, Dalton's atomic model, and the conservation of mass theory. This last mention has the most importance of the three due to the fact that this law intertwines chemistry with thermodynamics in an intercalated manner.

### 1.3.2 Enzymes

As early as the late 18th century and early 19th century, the digestion of meat by stomach secretions and the conversion of starch to sugars by plant extracts and saliva were known. However, the mechanism by which this occurred had not been identified.

In the 19th century, when studying the fermentation of sugar to alcohol by yeast, Louis Pasteur concluded that this fermentation was catalyzed by a vital force contained within the yeast cells called ferments, which he thought functioned only within living organisms. He wrote that "alcoholic fermentation is an act correlated with the life and organization of the yeast cells, not with the death or putrefaction of the cells."

In 1833 Anselme Payen discovered the first enzyme, diastase, and in 1878 German physiologist Wilhelm Kühne (1837–1900) coined the term enzyme, which comes from Greek "in leaven", to describe this process. The word enzyme was used later to refer to nonliving substances such as pepsin, and the word ferment was used to refer to chemical activity produced by living organisms.

In 1897 Eduard Buchner began to study the ability of yeast extracts to ferment sugar despite the absence of living yeast cells. In a series of experiments at the University of Berlin, he found that the sugar was fermented even when there were no living yeast cells in the mixture. He named the enzyme that brought about the fermentation of sucrose "zymase". In 1907 he received the Nobel Prize in Chemistry "for his biochemical research and his discovery of cell-free fermentation". Following Buchner's example; enzymes are usually named according to the reaction they carry out. Typically the suffix -ase is added to the name of the substrate (e.g., lactase is the enzyme that cleaves lactose) or the type of reaction (e.g., DNA polymerase forms DNA polymers).

Having shown that enzymes could function outside a living cell, the next step was to determine their biochemical nature. Many early workers noted that enzymatic activity was associated with proteins, but several scientists argued that proteins were merely carriers for the true enzymes and that proteins per se were incapable of catalysis. However, in 1926, James B. Sumner showed that the enzyme urease was a pure protein and crystallized it; Sumner did likewise for the enzyme catalase in 1937. The conclusion that pure proteins can be enzymes was definitively proved by Northrop and Stanley, who worked on the digestive enzymes pepsin (1930), trypsin, and chymotrypsin. These three scientists were awarded the 1946 Nobel Prize in Chemistry.

This discovery, that enzymes could be crystallized, meant that scientists eventually could solve their structures by x-ray crystallography. This was first done for lysozyme, an enzyme found in tears, saliva, and egg whites that digests the coating of some bacteria; the structure was solved by a group led by David Chilton Phillips and published in 1965. This high-resolution structure of lysozyme marked the beginning of the field of structural biology and the effort to understand how enzymes work at an atomic level of detail.

## 1.4 METABOLISM

### 1.4.1 Early Metabolic Interest

The history of the scientific study of metabolism spans 800 years. The earliest of all metabolic studies began during the early thirteenth century (1213-1288) by a Muslim scholar from Damascus named Ibn Al-Nafis. Al-Nafis stated in his most well-known work Theologus Autodidactus that "that body and all its parts are in a continuous state of dissolution and nourishment, so they are inevitably undergoing permanent change." Although al-Nafis was the first documented physician to have an interest in biochemical concepts, the first controlled experiments in human metabolism were published by Santorio Santorio in 1614 in his book Ars de statica medecina. This book describes how he weighed himself before and after eating, sleeping, working, sex, fasting, drinking, and excreting. He found that most of the food he took in was lost through what he called "insensible perspiration".

### 1.4.2 Metabolism: 20th Century - Present

One of the most prolific of these modern biochemists was Hans Krebs who made huge contributions to the study of metabolism. Krebs was a student of extremely important Otto Warburg, and wrote a biography of Warburg by that title in which he presents Warburg as being educated to do for biological chemistry what Fischer did for organic chemistry. Which he did. Krebs discovered the urea cycle and later, working with Hans Kornberg, the citric acid cycle and the glyoxylate cycle. These discoveries led to Krebs being awarded the Nobel Prize in physiology in 1953, which was shared with the German biochemist Fritz Albert Lipmann who also codiscovered the essential cofactor coenzyme A.

### 1.4.3 Glucose Absorption

In 1960, the biochemist Robert K. Crane revealed his discovery of the sodium-glucose cotransport as the mechanism for intestinal glucose absorption. This was the very first proposal of a coupling between the fluxes of an ion and a substrate that has been seen as sparking a revolution in biology. This discovery, however, would not have been possible if it were not for the discovery of the molecule glucose's structure and chemical makeup. These discoveries are largely attributed to the German chemist Emil Fischer who received the Nobel Prize in chemistry nearly 60 years earlier.

### 1.4.4 Glycolysis

Since metabolism focuses on the breaking down (catabolic processes) of molecules and the building of larger molecules from these particles (anabolic processes), the use of glucose and its involvement in the formation of adenosine triphosphate (ATP) is fundamental to this understanding. The most frequent type of glycolysis found in the body is the type that follows the Embden-Meyerhof-Parnas (EMP) Pathway, which was discovered by Gustav Embden, Otto Meyerhof, and Jakob Karol Parnas. These three men discovered that glycolysis is a strongly determinant process for the efficiency and production of the human body. The significance of the pathway shown in the adjacent image is that by identifying the individual steps in this process doctors and researchers are able to pinpoint sites of metabolic malfunctions such as pyruvate kinase deficiency that can lead to severe anemia. This is most important because cells, and therefore organisms, are not capable of surviving without proper functioning metabolic pathways.

### 1.4.5 Instrumental Advancements (20th Century)

Since then, biochemistry has advanced, especially since the mid-20th century, with the development of new techniques such as chromatography, X-ray diffraction, NMR spectroscopy, radioisotopic labelling, electron microscopy and molecular dynamics simulations. These techniques allowed for the discovery and detailed analysis of many molecules and metabolic pathways of the cell, such as glycolysis and the Krebs cycle. The example of an NMR instrument shows that some of these instruments, such as the HWB-NMR, can be very large in size and can cost anywhere from a few thousand dollars to millions of dollars.

### 1.4.6 Polymerase Chain Reaction

Polymerase chain reaction (PCR) is the primary gene amplification technique that has revolutionized modern biochemistry. Polymerase chain reaction was developed by Kary Mullis in 1983. There are four steps to a proper polymerase chain reaction: (1) denaturation (2) extension (3) insertion (of gene to be expressed) and finally (4) amplification of the inserted gene. These steps with simple illustrative examples of this process can be seen in the image below and to the right of this section. This technique allows for the copy of a single gene to be amplified into hundreds or even millions of copies and has become a cornerstone in the protocol for any biochemist that wishes to work with bacteria and gene expression. PCR is not only used for gene expression research but is also capable of aiding laboratories in diagnosing certain diseases such a lymphomas, some types of leukemia, and other malignant diseases that can sometimes puzzle doctors. Without polymerase chain reaction development, there are many advancements in the field of bacterial study and protein expression study that would not have come to fruition. The development of the theory and process of polymerase chain reaction is essential but the invention of the thermal cycler is equally as important because the process would not be possible without this instrument. This is yet another testament to the fact that the advancement of technology is just as crucial to sciences such as biochemistry as is the painstaking research that leads to the development of theoretical concepts.

## 1.5 FERMENTED FOODS

Fermented foods are defined as "foods or beverages produced through controlled microbial growth, and the conversion of food components through enzymatic action". Many foods have historically undergone fermentation, including meat and fish, dairy, vegetables, soybeans, other legumes, cereals and fruits. There are several variables in the fermentation process including the microorganisms, the nutritional ingredients and the environmental conditions, giving rise to thousands of different variations of fermented foods. Historically, food fermentation was performed as a method of preservation, as the generation of antimicrobial metabolites (e.g., organic acids, ethanol and bacteriocins) reduces the risk of contamination with pathogenic microorganisms. Fermentation is also used to enhance the organoleptic properties (e.g., taste and texture), with some foods, such as olives, being inedible without fermentation that removes bitter phenolic compounds.

There are two main methods through which foods are fermented. Firstly, foods can be fermented naturally, often referred to as "wild ferments" or "spontaneous ferments", whereby the microorganisms are present naturally in the raw food or processing environment, for example sauerkraut, kimchi, and certain fermented soy products. Secondly, foods can be fermented via the addition of starter cultures, known as "culture-dependent ferments", for example kefir, kombucha and natto. One method of performing a culture-dependent ferment is "backslopping", in which a small amount of a previously fermented batch is added to the raw food, for example sourdough bread. Starters used to initiate fermentation can be either natural (e.g., backslopping), or selected commercial starters to standardize the organoleptic characteristics of the final product.

Fermented foods hold a firm place in cuisine from almost every culture in the world. In the West, there has been a surge in popularity of fermented foods in recent years, major reasons including the proposed health benefits of fermented foods and surging interest in gastrointestinal health. There are several mechanisms through which fermented foods may exert beneficial effects in health and disease.

Firstly, they contain potentially probiotic microorganisms, such as lactic acid bacteria. In general, most fermented products have been found to contain at least 106 microbial cells per gram, with concentrations varying depending on several variables such as the product's region, age and time at which the products are analysed/consumed. The surrounding food matrix appears to play an important role in the survival of probiotic strains via its buffering and protective effect against gut conditions (e.g., low pH, bile acids). Indeed, a number of studies have shown that microorganisms from fermented foods can reach the gastrointestinal tract, this is likely to differ across products, and their presence in the gut appears to be transient. Nonetheless, these microorganisms may still have the potential to exert a physiological benefit in the gut, through competition with pathogenic bacteria and the production of immune-regulatory and neurogenic fermentation by-products. Secondly, fermentation-derived metabolites may exert health benefits. For example, lactic acid bacteria (relevant to both dairy and non-dairy fermented foods) generate bioactive peptides and polyamines with potential effects on cardiovascular, immune and metabolic health. Thirdly, fermentation may convert certain compounds to biologically active metabolites. For example, lactic acid bacteria can convert phenolic compounds (such as flavonoids) to biologically active metabolites. Fourthly, food components found in fermented foods, such as prebiotics and vitamins, may also exert health benefits. Lastly, fermentation can reduce toxins and anti-nutrients, for example, fermentation of soybeans may reduce phytic acid concentrations, and sourdough fermentation can reduce the content of fermentable carbohydrates (e.g., fermentable oligosaccharides, disaccharides, monosaccharides and polyols, FODMAPs), which may increase the tolerance of these products in patients with functional bowel disorders such as irritable bowel syndrome

This review aims to define and characterise common fermented foods, their mechanisms of action (including impact on the microbiota), and the evidence for effects on gastrointestinal health and disease in humans. The evidence for the effects of yoghurt and cheese on human health has been extensively reviewed elsewhere and, therefore, this review will focus specifically on kefir and the major non-dairy fermented foods: kombucha; sauerkraut; tempeh; natto; miso; kimchi; and sourdough bread (Table 1).

**Table 1: Description and microbial content of common fermented foods.**

| Name | Description | Region of Origin | Sourcen of Microorganisms | Microorganisms Identified in Final Product * |
|---|---|---|---|---|
| Kefir | Fermented milk beverage | Caucasus | Starter culture | Lactobacillus kefiri, Lactobacillus paracasei, Lactobacillus parabuchneri, Lactobacillus casei, Lactobacillus lactis, Lactococcus lactis, Acetobacter lovaniensis, Kluyveromyces Lactis, Saccharomyces cerevisiae |
| Kombucha | Fermented tea beverage | China | Starter culture | Komagataeibacter xylinus, Saccharomyces cerevisiae, Zygosaccharomyces bailii. Brettanomyces bruxellensis, Acetobacter pasteurianus, Acetobacter aceti, Saccharomyces cerevisiae, Zygosaccharomyces bailii, Brettanomyces bruxellensis, Acetobacter xylinum, Zygosaccharomyces spp., Acetobacter, Gluconacetobacter |
| Sauerkraut | Fermented cabbage | China | Spontaneous | Lactobacillus sakei, L. plantarum, Candidatus accumulibacter phosphatis, Thermatoga spp., Pseudomonas rhizosphaerae, L. hokkaidonensis, L. rhamnosus, Leuconostoc carnosum, Clostridium |

| Name | Description | Region of Origin | Sourcen of Microorganisms | Microorganisms Identified in Final Product * |
|---|---|---|---|---|
| | | | | saccharobutyrilicum, Rahnella aquatillis, Citrobacter freundii, Enterobacter cloacae, Bifidobacterium dentium, Enterococcus faecalis, Lactobacillus casei, Lactobacillus delbrueckii, Staphylococcus epidermidis, Lactobacillus curvatus, Lactobacillus brevis, Weissella confusa, Lactococcus lactis, Enterobacteriaceae, Leuconostoc spp., Yarrowia brassicae |
| Tempeh | Fermented boiled and dehulled soybeans | Indonesia | Starter culture (Rhizopus oligoporus) | Enterococcus faecium, Rhizopus oryzae, Rhizopus oligoporus, Mucor indicus, Mucor circinelloides, Geotrichum candidum, Aureobasidium pullulans, Alternaria alternata, Cladosporium oxysporum, Trichosporon beigelii, Clavispora lusitaniae, Candida maltosa, Candida intermedia, Yarrowia lipolytica, Lodderomyces elongisporus, Rhodotorula mucilaginosa, Candida sake, Hansenula fabiani, Candida tropicalis, Candida parapsilosis, Pichia membranefaciens, Rhodotorula rubra, Candida rugosa, Candida curvata, Hansenula anomola |
| Natto | Fermented soybean | Japan | Starter culture (Bacillus subtilis natto) | Data not available |

| Name | Description | Region of Origin | Sourcen of Microorganisms | Microorganisms Identified in Final Product * |
|---|---|---|---|---|
| Miso | Fermented soybean paste | Japan | Starter culture (Aspergillus oryzae) | Bacillus subtilis, Bacillus amyloliquefaciens, Staphylococcus gallinarum, Staphylococcus kloosii, Lactococcus sp. GM005 |
| Kimchi | Fermented vegetable dish | Korea | Spontaneous, Addedcommercially | Leuconostoc gasicomitatum, Leuconostoc gelidum, Leuconostoc mesenteroides, Weissella koreensis, Weissella confuse, Lactobacillus sakei, Lactobacillus plantarum, Lactobacillus curvatus, Trichosporon domesticum, Trichosporon loubieri, Saccharomyces unisporus, Pichia kluyveri |
| Sourdough bread | Bread made from longer ferment | Middle East and Europe | Spontaneous or backslopping | Data not available |

## 1.6 EXAMPLES OF FERMENTED FOODS

You may not know that some foods you eat often are actually. In general, it takes only a few ingredients to make some fermented foods at home, so they're easy to add to your diet.

Fermented foods include:

- Cheese
- Olives
- Yogurt
- Tempeh
- Miso
- Kimchi
- Sauerkraut
- Bread (if made with yeast)
- Buttermilk
- Kefir

- Kombucha
- Cider
- Wine
- Beer
- Pickles
- Sour cream
- Vinegar

### 1.6.1 Benefits of Fermented Foods

Fermented foods have historically been valued for their improved shelf life and unique taste, aroma, texture and appearance. They also allow us to consume otherwise inedible foods. For example, table olives must be fermented in order to remove their bitter-tasting phenolic compounds.

Many health benefits have been associated with fermented foods, including reduced risk of cardiovascular disease, high blood pressure, diabetes, obesity and inflammation. They have also been linked to better weight management, better mood and brain activity, increased bone health and better recovery after exercise. When looking at heart health, probiotics may help to decrease total and low-density lipoprotein (LDL) cholesterol however the evidence for this is still very limited.

One explanation for all of these effects is the production of bioactive peptides, vitamins and other compounds produced by the microorganisms involved in fermentation and have key roles in the body, such as blood health, nerve function and immunity.

It's important to remember that these health benefits are likely dependent on the type of fermented food and microorganisms involved. For example, yoghurt consumption has been associated with reduced risk of type 2 diabetes, while fermented milk that contains Lactobacillus helveticus has been associated with reduced muscle soreness.

### 1.6.2 Eating More Fermented Foods

Although fermented foods may sound fancy, the practice of fermentation is actually simple and affordable. It only requires a few ingredients and when done at home, can save you a lot of money, while adding variety, new flavours and interesting textures to your diet. Vegetables such as cabbage, beetroot, radish, turnip and carrots are some of the easiest foods to ferment at home, as the bacteria living on the surface does the fermenting for you.

Try making your own sauerkraut, kimchi and pickled seasonal vegetables, including prebiotic-rich foods such as onion and garlic to add flavour and extra health benefits.

- Kimchi – fermented cabbage recipe
- Pickled vegetables recipe

This is a great way to experience the benefits of fermented foods while including more vegetables in your diet.

Despite fermented foods being nutritious foods, there is no single food that improves our heart health – it is our overall diet.

Fermented foods are best eaten in the context of a heart-healthy eating pattern that emphasises vegetables, fruit, whole grains in place of refined grains, legumes, nuts, seeds and oily fish.

### 1.6.3 How are Foods Fermented?

Wild fermentation. In this process, also called spontaneous fermentation, the microorganisms that cause fermentation are already in the environment. For example, the microorganisms that make a sourdough starter — one of the ingredients in sourdough bread — come from the flour used to make it and often from the baker's hands.

Culture fermentation. This process uses a starter, like a sourdough starter, to ferment foods. Kombucha is another food that uses a starter – in this case, a symbiotic culture of bacteria and yeast (SCOBY).

### 1.6.4 Are Fermented Foods Good for You?

Research suggests that there are some health benefits to eating fermented foods. They may help the balance of helpful bacteria that live in your intestines and help you digest food. These microbes also help your immune system fight off harmful bacteria.

Fermented foods may be especially good after you've taken antibiotics. They can help restore the balance of good bacteria in your digestive system.

Some experts believe that an unbalanced microbiome leads to leaky gut syndrome, a condition in which your intestines leak nutrients into your bloodstream. It may contribute to a range of conditions including eczema and Alzheimer's disease. Balancing your gut biome with fermented foods may help strengthen the walls of your intestines.

When choosing fermented foods, keep in mind that those with live cultures are the most helpful for your gut biome. For example, there aren't usually live organisms in finished cheese, but there are in yogurt. Other examples of foods containing live cultures include:

- Kimchi
- Kefir
- Certain kinds of pickles

In addition to helping your gut biome, fermented foods can lower your risk of heart disease. They also may help reduce several key factors in heart disease like high blood pressure and obesity. They might lower your risk of diabetes and can help with inflammation. Preliminary studies also suggest that people who eat fermented foods may have lower "bad" LDL cholesterol levels.

### 1.6.5 Risks of Fermented Foods

If they have been prepared and handled properly, there are no documented risks to eating fermented foods. But if they're contaminated during or after the fermentation process, there is a risk of food poisoning.

There are some risks specific to drinking too much kombucha. Health experts recommend drinking no more than 12 ounces of the fermented tea per day. Symptoms of drinking too much kombucha include:

- Headache
- Nausea
- Stomachache
- Diarrhea
- Rash
- Sinus congestion
- Anxiety
- Muscle cramps

Another risk of kombucha is vinegar nematodes. These small worm-like creatures are not harmful if consumed but can make drinking kombucha unpleasant. If you have vinegar nematodes in your home-brewed kombucha, experts recommend throwing out the whole thing, including the SCOBY, and starting over from scratch.

## 1.7 SAFETY ANALYSIS OF FERMENTED FOODS

Fermentation is one of the oldest technologies used for food preservation. Over the centuries, it has evolved and been refined and diversified. Today, a variety of food products is derived from this technology in households, small-scale food industries as well as in large enterprises. Furthermore, fermentation is an affordable food preservation technology and of economic importance to developing countries.

Ethyl carbamate (EC), also known as urethane, is a chemical contaminant naturally formed in fermented foods during the fermentation process or during storage. Measurable levels of EC have been found in foods such as bread, soy sauce and yogurt, and in alcoholic beverages such as spirits, wine and beer. The health concern of EC in food is related to its carcinogenic potential. Ethyl carbamate has a history of use in industry, medicine and veterinary applications. Its use in human medicine was later banned due to toxicological concerns and lack of efficacy.

The levels of EC present in different alcoholic beverages and fermented foods have been extensively reviewed in overseas countries. Studies demonstrated that foods produced at least partially by yeast fermentation seemed to be much more likely to contain EC than those fermented by lactic acid bacteria, acetic acid bacteria and moulds, while the non-yeast fermented products (e.g. cheese) usually contained low to non-detectable levels of EC.

There is currently no international standard for the maximum allowable level of EC in foods. However, some countries have established maximum levels of EC in alcoholic beverages. Canada was the first country to introduce maximum levels for ethyl carbamate in a variety of alcoholic beverages, from 30 μg/L for wine to 400 μg/L for fruit brandies. The US has voluntary target levels of EC for her national food products, and has notified all countries exporting wines to the US that they must develop programs to meet these target levels. There are currently no harmonized maximum levels of EC in foods in the European Union, although some member states have recommended maximum EC levels in alcoholic beverages. Recently, Korea has also established a maximum EC level of 30 μg/L in wine.

At present, there is no subsidiary legislation governing the maximum level of EC in fermented foods and alcoholic beverages in Hong Kong. While previous studies carried out overseas mainly focused on food items of the Western diet, no comprehensive studies on the level of contamination by EC in commonly consumed local fermented foods and beverages have been conducted. The health risk associated with dietary exposure to EC upon consumption of fermented foods and beverages for the local population is not clear.

It is important that a risk assessment study on EC be conducted, particularly on the local fermented foods such as preserved vegetables, fermented soy products, local condiments and sauces, and alcoholic beverages including Chinese fermented wines. The study would be of value to both the local and international community, for providing first-hand information on the level of contamination of EC in the commonly consumed local fermented foods in Hong Kong, for assessment of the associated health risk of dietary exposure to EC of the local population, and for submission of the EC contamination data to the JECFA database.

### 1.7.1 Hazard Identification

Ethyl carbamate is the ester of carbamic acid. It is naturally formed in fermented foods mainly as a byproduct of fermentation, primarily from the reaction of alcohol (ethanol) with urea and its break-down products. Cyanate is probably the ultimate precursor in most cases, reacting with ethanol to form the carbamate ester.

Ethyl carbamate can be formed from various substances derived from fermented foods and beverages, including urea, hydrogen cyanide, citrulline and other N-carbamyl compounds. For example, in stone fruit spirits, cyanogenic glycosides from the stones may be degraded through enzymatic action to cyanide which is then oxidised to cyanate and reacts with ethanol to form EC. Another example is the formation of urea from the degradation of arginine by yeast fermentation, which is then broken down to isocyanate and reacts with ethanol to form EC. The chemical reaction between urea and ethanol has been found to accelerate exponentially with increase in temperature. Besides temperature, the presence of light and duration of storage are the other two key factors influencing the formation of ethyl carbamate in fermented foods.

The main source of dietary exposure to EC in human population is through the consumption of fermented foods and beverages containing EC. Alcoholic beverage is the known main contributor to EC exposure.

### 1.7.2 Hazard Characterisation

The Joint Food and Agriculture Organization/World Health Organization Expert Committee on Food Additives (JECFA) evaluated EC in 2005 and concluded that the chemical was genotoxic and a multisite carcinogen in all animal species tested and a potential carcinogen in human.

In 2007, the International Agency for Research on Cancer (IARC) re-classified EC as "probably carcinogenic to humans".

## 1.8 KINETICS AND METABOLISM

Ethyl carbamate is absorbed rapidly and almost completely from the gastrointestinal tract and is evenly distributed throughout the body. It is also eliminated quickly. In mice, more than 90% of the absorbed EC is eliminated as carbon dioxide within 6 hours.

Ethyl carbamate is metabolised via three main metabolic pathways, namely, hydrolysis, hydroxylation and side-chain oxidation. Vinyl carbamate epoxide, one of the metabolic intermediates of EC via side-chain oxidation, is considered the main metabolite responsible for its carcinogenicity. It binds covalently to nucleic acids and proteins, resulting in the formation of adducts including those that have been shown to induce base-pair substitutions in DNA from tumour tissue. Vinyl carbamate epoxide is further metabolised to carbon dioxide and ammonia and excreted in the urine.

### 1.8.1 Developmental Toxicity

While no multigeneration studies that meet currently accepted standard protocols are available, dose-related increases in skeletal anomalies have been observed in mice given single oral doses of EC at 300 – 1000 mg/kg bw on day 11 of gestation. Increases in external malformations and skeletal abnormalities have also been reported in rats given EC at daily doses of 1000 mg/kg bw by gavage for 1, 2, 6 or 7 consecutive days during the period of organogenesis (days 6 – 13 of gestation).

Treatment of female mice with single or multiple doses of EC during gestation or lactation was found to increase the incidence or multiplicity of tumours in the adult offspring compared with untreated controls.

### 1.8.2 Mutagenicity and Genotoxicity

Ethyl carbamate is considered a genotoxic mutagen in experimental animals. It has been tested in a large number of genotoxicity studies in vitro and in vivo. The results of assays for point mutations were uniformly negative for mouse lymphoma cells, while assays in bacterial, yeast and other types of mammalian cells produced variable results. Results of assays in somatic cells in vivo (including tests for induction of chromosomal aberrations, micronucleus formation and sister chromatid exchange) were almost uniformly positive. The assay for micronucleus formation in mice showed the strongest positive response, and co-administration of ethanol delayed, but did not inhibit the genotoxicity of EC in this assay. The results of assays for dominant lethal mutations or specific locus tests in mice, given EC either by intraperitoneal injection or in drinking-water, revealed no evidence for genotoxicity in mammalian germ cells in vivo.

### 1.8.3 Observation in Human

There are at present very few EC data in humans. JECFA considered in its 2005 evaluation that the human data available were not of a quality that could be used for hazard characterisation.Risk assessment for ethyl carbamate

JECFA evaluated EC in its 64th meeting in 2005 and concluded that EC was genotoxic and a multisite carcinogen in all animal species tested and a potential carcinogen in human.

The pivotal toxicological study for risk assessment was a carcinogenicity study of EC in mice. The increased incidence of lung tumour (alveolar and bronchiolar adenoma or carcinoma) was considered the critical toxicological end-point of concern. The associated dose-response data were analysed and the Benchmark Dose Lower Confidence Limit 10% (BMDL10) for EC was determined.

The BMDL10 represents the lower bound of a 95% confidence interval on a benchmark dose corresponding to a 10% tumour incidence (i.e. 10% extra risk of tumours) in experimental animals. The EC margin of exposure (MOE), defined as the ratio of BMDL10 to the estimated dietary intake of EC in humans, was then calculated and used by the Committee as an index for the assessment of the potential health risk posed by EC to humans.

When the estimated intake of EC in foods excluding alcoholic beverages 14 (15 ng/kg bw/day) was compared with the lower end (0.3 mg/kg bw/day) of the range of BMDL10 values (0.3 to 0.5 mg/kg bw/day) for alveolar / bronchiolar neoplasms in mice, a MOE of 20,000 was estimated. With the inclusion of alcoholic beverages in the estimated intake of EC (80 ng/kg bw/day), the resulting MOE was only 3,800. The Committee concluded that exposure to EC in foods other than alcoholic beverages would be of low concern, but the estimated health risk posed by total dietary exposure to EC (in all foods and beverages) would be of potential concern and therefore mitigation measures to reduce concentrations of EC in some alcoholic beverages should be continued.

## 1.9 FERMENTATION MICROORGANISMS AND FLAVOR CHANGES IN FERMENTED FOODS

Lactic acid bacteria influence the flavor of fermented foods in a variety of ways. In many cases, the most obvious change in a lactic acid fermentation is the production of acid and lowering pH thatresults in an increase in sourness. Since most of the acid produced in fermentations will be produced by the metabolism of sugars, sweetness will likely decrease as sourness increases. The production of volatile flavor components tends to be the first mechanism considered for the development of flavor specific to a particular fermented food. In addition to this direct mechanism, however, there are less direct ways in which fermentation microorganisms affect flavor. Lowering the pH in lactic acid fermentations may reduce the activity or completely inactivate enzymes in the plant that generate either flavor components or flavor precursor compounds.

Finally, the fermentation microorganisms may directly metabolize precursor flavor compounds or flavor components themselves.

### 1.9.1 Prevention of Flavor Formation

Purge-and-trap analysis of the volatile components found in cucumber slurries before and after cucumbers were fermented in a 2% reduced-salt brine. Comparison of volatile components before and after fermentation led to the conclusion that the major effect of the fermentation on flavor volatiles was to prevent enzymatic formation of E, Z-2, 6-nonadienal and 2-nonenal by enzymes present in cucumbers. These aldehydes are the major compounds responsible for fresh cucumber flavor. However, a few days into cucumber fermentation, the pH drops low enough to inactivate the enzymes which form these compounds when cucumber tissue is disrupted. Among the volatile components identified in the fermented cucumbers, only benzaldehyde, ethyl benzene, and o-xylene were not observed in fresh cucumber slurries. The lack of the flavor impact of volatile aldehydes is certainly the major effect of the fermentation on flavor.

More recently, Marsili and Miller (2000) found a low volatility flavor impact compound in fermented pickled cucumber brines. Addition of saturating salt to brine samples heated to 50 °C, sampling with an SPME (solid-phase microextraction) fiber and followed by GC-olfactometry led to recognition of a compound with an odor close to that of the fermentation brine. The compound with a fermentation brine odor was identified as trans-4-hexenoic acid. They also tentatively identified the presence of cis-4-hexenoic acid. In a reconstitution experiment, a solution that contained 25 ppm trans- 4-hexenoic acid, 10 ppm phenyl ethyl alcohol, 0.65% lactic acid, 0.05% acetic acid, and 8% NaCl had an odor very similar to that of brine from fermented cucumbers. The concentrations of the lactic acid, acetic acid, and NaCl are reasonable for commercial brines after the completion of fermentation. The addition of phenyl ethyl alcohol gave only a small improvement in the odor match. Thus, the trans-4-hexenoic acid was the key component in the simulated brine solution. Unfortunately, the origin of trans-4-hexenoic acid in fermentation brines is not known.

### 1.9.2 Volatile Odor Changes Caused by Lactic Acid Fermentation

In contrast to the cucumber fermentation, where most volatile components did not change substantially, Czerny and Schieberle (2002) observed a number of changes in odorants when whole meal wheat flour was fermented by a commercial sourdough starter culture. Using aroma extract dilution analysis, they identified 14 compounds that had substantial odor intensity in fermented sourdough. All of these odorants were present in dough before and after fermentation, so the fermentation did not result in formation of compounds not present in the flour, nor did fermentation completely remove odor components. However, most of the compounds changed during fermentation, with changes ranging from 7-fold decreases to 9-fold increases in concentration.

### 1.9.3 Flavor Precursor Formation Caused by Lactic Acid Fermentation

When sourdough bread is baked, one of the characteristic odor compounds generated in the baking process is 2-acetyl pyrroline, which has a roasted, popcorn-like odor, with an odor threshold of only 20 ppt. The precursor of 2-acetyl pyrroline during baking is ornithine. Ornithine is synthesized from free arginine by heterofermentative lactic acid bacteria. De Angelis and others (2002) have shown that Lactobacillus sanfranciscensis has the enzymes arginine deiminase and ornithine transcarbamoylase, which can convert arginine to ornithine. However, L. sanfranciscen sis only increased the ornithine concentrations during dough fermentation when additional arginine was added. Thiele and Vogel (2002) reported that ornithine accumulated during sourdough fermentation only when Lactobacillus pontis was used as the starter culture. No increase in ornithine was found when the starter was L. sanfranciscensis. This result indicates the difficulty that can be encountered between showing that a fermentative organism can carry out a reaction and showing that, in the complex conditions existing in a food fermentation, the expected transformation will, in fact, occur.

While 2-acetyl-pyrroline is apparently formed from ornithine by chemical reactions in bread baking, it may be formed by lactic acid bacteria in wine. It is one of the compounds associated with the highly objectionable 'mousy' off-flavor in wine. Costello and Henschke (2002) showed that Lactobacillus hilgardii, a lactic acid bacterium that occurs in wine, could produce the compound from ornithine in a defined medium with a high cell density. Formation of 2-acety l-pyrroline also required fructose as a fermentable sugar, ethanol, and iron ($Fe^{+2}$) in the medium.

# 2

# FERMENTATION TECHNOLOGY

## 2.1 INTRODUCTION

Fermentation technology is a field which involves the use of microorganisms and enzymes for production of compounds which have application in the energy, material, pharmaceutical, chemical and the food industry. Though fermentation processes are used for generations for the requirement for sustainable production of materials and energy is demanding creation and advancement of novel fermentation processes.

Efforts are directed both to the advancement of cell factories and enzymes as well as of design of new processes, concepts and technologies for fermentation process. Through fermentation, we can produce enzymes for industrial purposes. Process of Fermentation includes the use of microorganisms, like yeast and bacteria for the production of enzymes.

Mainly, there are two methods of fermentation which are used to produce enzymes. First is submerged fermentation and second is solid-state fermentation. In Submerged fermentation, the production of enzymes is done by microorganisms in a liquid nutrient media. Whereas in Solid- fermentation is carried out by cultivation of microorganisms and production of enzyme is done on a solid substrate. Compounds containing carbon in or on the substrate are busted down by the micro organisms thus producing the enzymes either extracellular or intracellular. The enzymes are isolated by various methods such as centrifugation, and for extracellular produced enzymes and lying of cells for intracellular enzymes.

Various industries rely on enzymes for the production of their enzyme dependent products. Industries which are used for enzymes generated by fermentation are the wine making, brewing, cheese making and baking.

Fermentation Technology is the longest-run course in the MIT Professional Education catalog. It has been offered continuously for more than 50 years. This course emphasizes the application of biological and engineering principles to problems involving microbial, mammalian, and biological/biochemical systems. The aim of the course is to review fundamentals and provide an up-to-date account of current knowledge in biological and biochemical technology. The lectures will emphasize and place perspectives on biological systems with industrial practices.

This course has made some major additions, modifications, and revisions in the course topics and course contents over the past few years. In recognition of the increasing number of attendees from non-pharmaceutical industries, the instructors are balancing the course to provide equal emphasis on mammalian and microbial technologies. More than half of the lecturers are currently working in industry or have industrial experience.

## 2.2 FERMENTATION METHODOLOGY

Fermentation process is carried out in a container called the fermentor or bioreactor. The design and nature of the fermentor varies depending upon the type of fermentation carried out. Invariably all the fermentors have facilities to measure some of the fermentation parameters like temperature, pressure, pH, elapsed fermentation time, liquid level, mass etc.

The different types of fermentors are:

(1) External recycle airlift fermentor—for producing bacterial biomass, with methanol as substrate.

(2) Internal recycle airlift fermentor—for producing yeast with oil as substrate.

(3) Tubular tower fermentor—Used for making beer, wine, vinegar etc.

(4) Nathan fermentor—used in brewing industry.

(5) Stirred fermentor—used for making antibiotics.

### 2.2.1 Types of Fermentation Processes

There are three different process of fermentation viz.:

(1) Batch fermentation

(2) Feb-batch fermentation and

(3) Continuous culture.

**Batch Fermentation:** This term is attributed to that type of fermentation wherein there is change in culture medium, number of microorganisms and the amount of the product produced (i.e. the metabolite or target protein). In batch fermentation six phases of the microbial growth are seen.

**(a) Lag Phase:** Immediately after inoculation, there is no increase in the numbers of the microbial cells for some time and this period is called lag phase. This is in order that the organisms adjust to the new environment they are inoculated into.

**(b) Acceleration Phase:** The period when the cells just start increasing in numbers is known as acceleration phase.

**(c) Log Phase:** This is the time period when the cell numbers steadily increase.

**(d) Deceleration Phase:** The duration when the steady growth declines.

**(e) Stationary Phase:** The period where there is no change in the microbial cell number is the stationary phase. This phase is attained due to depletion of carbon source or accumulation of the end products.

**(f) Death Phase:** The period in which the cell numbers decrease steadily is the death phase. This is due to death of the cells because of cessation of metabolic activity and depletion of energy re¬sources. Depending upon the product required the different phases of the cell growth are maintained. For microbial mass the log phase is preferred. For production of secondary metabolites i.e. antibiotics, the stationary phase is preferred.

**Feb-batch Fermentation:** In this type of fermentation, freshly prepared culture media is added at regular intervals without removing the culture fluid. This increases the volume of the fermentation culture. This type of fermentation is used for production of proteins from recombinant microorganisms.

**Continuous Fermentation:** In this type of fermentation the products are removed continuously along with the cells and the same is replenished with the cell girth and addition of fresh culture media. This results in a steady or constant volume of the contents of the fermentor. This type of fermentation is used for the production of single cell protein (S.S.P), antibiotics and organic solvents.

## 2.2.2 Procedure of Fermentation

(a)Depending upon the type of product required, a particular bioreactor is selected.

(b) A suitable substrate in liquid media is added at a specific temperature, pH and then diluted.

(c) The organism (microbe, animal/plant cell, sub-cellular organelle or enzyme) is added to it.

(d) Then it is incubated at a specific temperature for the specified time.

(e) The incubation may either be aerobic or anaerobic.

  i. Aerobic conditions are created by bubbling oxygen through the medium.

  ii. Anaerobic conditions are created by using closed vessels, wherein oxygen cannot diffuse into the media and the oxygen present just above is replaced by carbon dioxide released.

(f) After the specified time interval, the products are removed, as some of the products are toxic to the growing cell or at least inhibitory to their growth. The organisms are re-circulated. The process of removal of the products is called downstream processing.

### *Example of Batch Fermentation Technology Using Tubular Tower Fermentor*

The simplest and the most commonly used fermentation technology is the preparation of curd from milk.

The details of the technology are as under:

**The fermentor or bioreactor:** Mud pot or steel cooking vessel or a dish or a cup

**The substrate:** Milk

**The specific temperature:** 37°C

**The organism:** Preformed curd (containing the microbe Lactobacillus cecai)

**The incubation:** At 37° C for 6-8 hours

**The aeration:** The process is anaerobic

**The process:** Preformed curd has microbes that utilize lactose present in the milk. Lactose is hydrolysed into glucose and galactose. Galactose is converted to glucose. Glucose is broken down to lactic acid by glycolytic pathway. The lactic acid produced, lowers the pH of milk from 6.6 to 4.5. The isoelectric pH of milk protein-casein is 4.5. At this pH, casein precipitates forming fine micelles in the milk thereby curdling it.

Milk is warmed before adding it to the fermentor so as to maintain the temperature required for the growth of the microbes. The container/fermentor is not disturbed (no stirring is taken up) so that the precipitation is uniform.

The elapsed time i.e. the time required to form the curd is crucial; it depends upon the atmospheric temperature. During summer the curd is formed within 4-6 hours, whereas during rainy season it takes 6-8 hours and in winter it takes almost 8-12 hours. The process is anaerobic hence it is better to keep the vessel closed. Though it cannot be air tight but still the surface of the milk that contains fat prevents air from penetrating in the liquid, furthermore the metabolizing microbes replace the oxygen with carbon dioxide released.

### *Example of Continuous Fermentation Technology Using Stirred Fermentor*

Another common fermentation technology is the process of food digestion.

The details of the technology are as under:

**The fermentor or bioreactor:** Stomach (G.I. tract)

**The substrate:** Food

**The specific temperature:** 37°C

**The organism:** Enzymes, microbes, acids etc.

**The Incubation:** At 37°C for 3-4 hours

**The aeration:** The process is anaerobic

**The process:** The food eaten by us is stored in the stomach where HC1 and some enzymes are secreted that convert food into chyme (semi solid). The food stays for 3-4 hours at 37° C. The digested food is absorbed in the blood and the undigested food is excreted. This is a continuous fermentation because the substrate (food) is continuously added and the products (digested/undigested material) are continuously removed. Stomach-the fermentor is a stirred type i.e. the peristaltic movement of the gastro-intestinal tract mixes the food.

## 2.2 PRINCIPLE OF FERMENTATION

- The main principle of fermentation is to derive energy from carbohydrates in the absence of oxygen.
- Glucose is first partially oxidized to pyruvate by glycolysis.
- Then pyruvate is converted to alcohol or acid along with regeneration of NAD+ which can take part in glycolysis to produce more ATP.
- Fermentation yields only about 5% of the energy obtained by aerobic respiration.

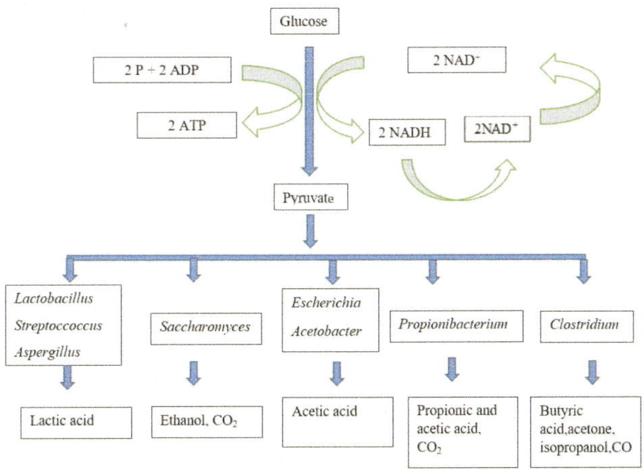


**Fig. 1. Flowchart: Generalized pathways for the production of some fermentation end products from glucose by various organisms.**

- Fermentation is an anaerobic biochemical process that is used for the production of energy from the partial oxidation of glucose or other carbon sources.
- The oxidation of the substrate, which occurs through the Embden–Meyerhoff (EMP) or Entner–Doudoroff(ED) pathways, results in the production of pyruvate, ATP, and NAD (P) H.
- In the absence of external electron acceptors, the pyruvate undergoes reduction with the regeneration of NAD+(P).
- This step is essential for the fermentation process to progress and it leads to the production of products (ethanol and organic acids).

- ATP is the main product of fermentation, and it is generated by phosphorylation at the substrate level.
- NADH is then re-oxidized, reverting to NAD+ in the second phase of fermentation that reduces pyruvate to the fermentation product, such as ethanol and lactate.
- For example, in the fermentation of glucose by Streptococcus lactis, the pyruvate is converted to lactic acid to reform NAD+ coenzymes so two ATP molecules are produced.
- In yeasts like Saccharomyces, when pyruvate is converted to ethyl alcohol (ethanol), NAD+ is reformed.

## 2.4 CATEGORIES OF FERMENTATION TECHNOLOGY

Fermentation technology can be grouped into four major categories viz.

**1. Microbial biomass production:** Microbial cells (biomass) are grown commercially as continuous culture on a large scale ($1500/m^3$). The microbial cells including algae, bacteria, yeasts, moulds and mushrooms are dried and used as a good source of a complete protein called 'single cell protein (SCP)' which serves as human food or animal feed. The incubation of wheat flour for preparation of 'tandoori roti' and rice flour for 'dosa' are good examples for the production of single cell protein.

These two food stuffs or the common food are usually lacking or not having sufficient quantities of lysine and methionine amino acids. Single cell protein produced by different sources is a rich source of different essential amino acids, thereby supplementing the amino acid lacking in a particular food. The incubation of the foods results in the growth of the microbial biomass, producing the SCP.

SCP produced by some microorganisms is lysine rich whereas those produced by others are methionine rich. The substrates used by the microbes producing the single cell protein range from carbohydrates to hydrocarbons and petrochemicals. Other organisms use the gases—methane, ethane, propane, n-butane etc as substrate for fermentation.

**SCP producing organisms and their substrates**

| Member | Substrate used |
|---|---|
| *Chlorella pyrenoidesa* | $CO_2$ (0.7% in air), fluorescent light |
| *Scenedesmus acutus* | $CO_2$, sunlight |
| *Spirulina maxima* | $CO_2$ (0.5%), $NaHCO_3$, sunlight |
| *Achromobacter deluacuate* | Diesel oil |
| *Nocardia sps.* | n-Alkanes |
| *Pseudomonas sps.* | Fuel oil |
| *Candida tropicalis* | Molasses |
| *Trichoderma sps.* | Coffee wastes |

Methylococcus capsulatus, Saccharomyces cerevisiae and Aspergillus.

**2. Microbial metabolites:** During the metabolism of microbial cells a number of compounds are produced and many are secreted out of the cell, which can be easily extracted and are very useful to man and animals. Therefore fermentation by microbial cells is carried out on an industrial scale, in order to get various metabolites.

The metabolites produced by the microbes can be grouped into two categories:

(a) Primary metabo¬lites and

(b) Secondary metabolites.

**(a) Primary metabolites:** Metabolites which are produced by the metabolism required for the main¬tenance of the minimum life process of a microbe are known as primary metabolites. The primary metabolites are produced in abundance at an early stage of growth. Examples of primary metabolites are ethanol, citric, acid, glutamic acid, lactic acid, acetic acid, acetone, formic acid, butanol, propionic acid, dihydroxy-acetone, glycerol etc. These metabolites are produced by fermentation technology applying different microbes under varying conditions of fermentation.

**(b) Secondary metabolites:** Secondary metabolites are those metabolites, which are not produced directly by the metabolism required for the vital life process of microbes, instead are produced by some specialized metabolic process. However most of the secondary metabolites are derived from the primary metabolites. The secondary metabolites include the antibiotics, alkaloids, toxic pigments, vitamins etc.

**Antibiotics:** An antibiotic is a substance produced by a microorganism which can inhibit growth or completely destroy other microorganisms. The antibiotics are not synthesized by the microorganism by a single gene, instead a set of 10-20 genes takes part in the synthesis of an antibiotic. The genes for antibiotic are mostly located on the plasmid, however in some microbes they are found on the chromosomal DNA.

About 45000 different antibiotics have been discovered so far, but only about 100 are used for human treatment, because only those antibiotics can be used which are toxic to the invading microbe but non-toxic for the human.

Some of the antibiotics and the microbes which produce them are:

Amphotericin-B—Streptomyces nodosus Chloramphenicol—Streptomyces venezuelae, Erythromycin—S. eythirus, Gentamycin—Micromonospora purpurea, Gramicidin—Bacillus brevis, Penicillin—Penicillin chryosogen Streptomycin—S. grises, Tetracycline—S. aureofacins.

These microbes are grown under suitable fermentation conditions so as to get the desired antibiotic. Mutant microbes are used for industrial production of antibiotics.

**3. Microbial enzymes:** When microbes are cultured, they secrete some enzymes into the media and these enzymes are extracted and widely used in several industries like detergent, food processing, brewing and pharmaceutical. They are also used for diagnostic, scientific and analytical purposes. Biotechnological methods are used to engineer microbial cells so as to induce them to produce enzymes like renin by E. coli and amylases by Bacillus stearothermophillis.

These enzymes are generally bound to matrix or in some manner retained in the reactor to be reused and hence these are called immobilized enzymes. In some cases the microorganism producing enzyme is immobilized. Some of the enzymes produced by fermenting microbes are—Glucose oxidase, protease glucoamylase, amylase, glucose isomerase, rennin, pectinase, superoxide dismutase, cellulase, invertase, lactase and lipase.

Some thermophile bacteria produce enzymes that are thermo-stable and which can be used in industrial processes at high temperatures, ex. glyceraldehyde-3-phosphate dehydrogenase, phosphofructo-kinase, alcohol dehydrogenase, superoxide dismutase and restriction endonucleases, which are produced by Bacillus stearothermophillis thermoysia. Further, genes for thermophilic enzymes are introduced into E. coli, which is cultured for producing the thermophilic enzymes.

**4. Bioconversion, biotransformation or modification of the substrate:** The fermenting microbes have got the capacity to convert an added substrate into some more valuable product, ex. conversion of ethanol to acetic acid (vinegar), isopropanol to acetone, glucose to gluconic acid, sorbitol to sorbose, sterols to steroids.

Among all these bioconversions, the production of steroids is the most widely applied fermentation biotechnology for the conversion of sterols into steroids, like cortisone, hydroxycortisone, prednisone, dexamethasone, testosterone, estradiol etc. Hitherto, steroids were produced by chemical methods, which were laborious and costly.

For instance the chemical synthesis of cortisone required 37 steps under extreme conditions. One of the steps is introduction of oxygen at position 11 in the steroid nucleus. The microbe Rhizopus arrhizus being capable of hydroxylating progesterone (a steroid) at position 11 is used in the fermentation to produce progesterone an intermediate in the synthesis of cortisone. Thus, this reduced the chemical synthesis steps from 37 to 11 and all under normal conditions, thereby making it economical and easy. Steroids are used as anti-inflammatory agents, contraceptives, treating hormonal insufficiency, allergy, skin diseases etc.

**Downstream processing (DSP):** The method by which the products of fermentation are recovered and separated is known as downstream processing. This forms the major (about 85%) portion of the complete fermentation technology. There are various methods by which DSP is carried out.

First of all, the broth is conditioned i.e. the cells are aggregated and form large clumps, which makes the separation easier. The conditioning is done by heating, freezing, pH change, antigen-antibody reactions etc. Then the conditioned broth is used for the separation of the constituents for which techniques like sedimen-tation, floatation, filtration, ultra-filtration, centrifugation and micro-filtration are applied.

## 2.5 DEVELOPMENT IN FERMENTATION TECHNOLOGY

Microbes used for fermentation process of food products develop flavor. Indeed, they also generate bioactive components that enhance the biofunctionality of food products and these foods called functional foods. Increasing prevalence of various chronic diseases i.e. obesity, diabetes, cardiovascular diseases and cancer that are associated with poor food habits, demands to develop new microbes for fermented food production that can enhance biofunctionality of foods against these life threatening health ailments.

Fermentation process is a process which involves the conversion of large molecules to small molecules or molecular oxidation/ reduction mechanisms mediated by selected micro-organisms. The fermentation technology depends on the microbial components and produces different molecules from small laboratory scale to large industrial scale. During this process various bioactive molecules also be produced, that can exert excellent health benefits for consumers. For example various bacteria and yeast have been used to produce conjugated linolenic acid (CLA), folate, vitamins and other bioactive components. Probiotic organisms are one of the important groups of such micro-organisms that participates in fermentation process of various food products i.e. dairy products. Probiotics are defined as live micro-organisms that exert health beneficial effects to host when administered on sufficient amount. Probiotic have been considered Generally Recognized As Safe for human consumption and known to exhibit various health beneficial effects i.e. anti-cancer, immunomodulatory, anti-oxidant, cardio protective and anti-obese/diabetic. There are various mechanisms have been proposed for the bioactivity of the probiotics on human health, and one of them is the probiotics mediated production of bioactive components during fermentation process of foods (Figure 2).

**Fig. 2. Development of fermented foods for human health: a projected future approach.**

Various chronic disease i.e. cardiovascular diseases, obesity, diabetes, cancer and auto-immune diseases are closely related to the food habits and quality of food. High energy diet/ fat consumption is one of the importance risk factor associated with obesity, diabetes and cancer. Fat is a basic unit of our cellular structure, therefore it is one of the essential components of our diet. Hence not all fats are detrimental for human health and can't be cut-down completely from diet. Excess saturated fatty acids are not considered good for health whilst unsaturated fatty acids are comparatively health beneficial. High fat foods and raw dairy products are one of the major sources for saturated fatty acids and if these products can be enriched with micro-organisms that can convert saturated fat to unsaturated fat should be one of the promising areas for fermentation technology. Although there are still a number of limitations in incorporating microbes in all kind of high fat foods, common food products such as ice cream, muffins, donuts, dairy products and other fermented foods can be enriched with such microbes that are able to convert most of the saturated fat into unsaturated fat and other bioactive fats i.e. CLA. Another kind of bioactive components produced by microbes during fermentation process of foods are bioactive peptides. Various probiotics and other microbes have been known for production of bioactive peptides from milk proteins, wheat protein and various other plant and meat proteins, that exert various health beneficial effects i.e. anti-hypertensive, immunomodulatory and anti-cancer. Supplementation of such microbes during the production process of foods that involve processing steps of fermentation can give promising fermented products or functional foods that can exhibit important biological activities to the consumers.

In addition various microbes have been reported to produces short chain fatty acids from dietary fats, carbohydrates, proteins and fibers i.e. butyrate that provides vital anti-obese and cardio protective effects. In this regard, pilot ongoing studies targeting specific cardiovascular-metabolic proinflammatory polymorphisms clusters are preliminarily proving to be useful predictor to help physicians identifying those subjects with a more consistent clinical response to an integrated gut ecology and nutritional-targeted intervention. If dairy products and other fat enriched foods can be supplemented with such micro-organisms during development, then fermentation process for development for functional foods can be of high importance in fermentation technology and food industry. The limitation in this direction might be the incorporation of these microbes may change in taste and flavor of the final food products.

However, these limitations can also be omitted by carefully standardizing the amount of microbes, time and temperature of fermentation process. In addition, the original taste and flavor can also be maintained by incorporating the bioactive microbes with original cultures that are responsible for flavor and taste i.e. yogurt cultures and cheese cultures with probiotic microbes. Also, original cultures can be genetically modified to incorporate the bioactive molecule production metabolic pathways along with intact flavor producing metabolic pathways; can be one of attractive areas of intensive research. Using such microbes will provide double edged beneficial effects i.e. microbes will produce bioactive components in foods that will be absorbed immediately after food consumption, and will also deliver such beneficial microbes in consumer's gastrointestinal tract, that further can produce such bioactive components into the gut and made them available to the host for long term basis. This is an active area of research and much more intensive research efforts should be directed in production of such beneficial microbes and functional foods that can ameliorate the global crisis of most prevalent chronic diseases i.e. heart diseases, obesity, diabetes, cancer and immune dysfunctions.

Fundamental research in enzymology have remarkably developed during the past thirty years. However, the actual application of enzymes has been rather limited for several technical and economical reasons. Such limiting factors are now being overcome year after year. Immobilized enzyme and cell technology might be a kind of break-through in this field. From the industrial view-point, enzyme technology is now coming into its real developmental stage.

Enzyme technology has recently been finding several directions for its development. These directions may be classified as follows:

1. New industrial catalysts,
2. Cools for food production and processing,
3. Pharmaceutical uses,
4. Analytical and measurement tools,
5. Aids for screening new physiologically active substances, and
6. Aids for creation of new sources of energy and raw materials.

## 2.5.1 Recent Developments in Biotechnology

Category Examples:

1. **Medicine**—Production of antibiotics, steroids, monoclonal antibodies, vaccines,gene therapy, recombinant DNA technology drugs and improving diagnosis by enzymes and enzyme sensors.
2. **Agriculture**—Plant tissue culture, protoplast fusion, introduction of foreign genesinto plants and nitrogen fixation.
3. **Chemicals** — Organic acids (citric, gluconic) and mineral extraction.
4. **Environment**—Improvementofwastetreatment,replacementofchemicalinsecticides by biological ones and biodegradation of xenobiotics.
5. **Food**—Single cell protein (SCP), use of enzymes in food processing andfood preservation.
6. **Industry**—Use of enzymes in detergent industry, textile and energy production.

### 2.5.2 Future Prospects for the Development of Biotechnology

Biotechnology is expected to make an important contribution to the quality of life through a wide range of goods and services in many areas for example:

1. Consumer chemicals: adhesives, detergents, dyes, fibers, flavors, gelling and thickening agents, gums, perfumes, pigments, plastics,........ etc.
2. Health care (diagnosis and treatment).
3. Analytical chemical tools.
4. Feed stocks (agriculture and animal).
5. Energy sources.
6. Environment control (air, water and soil)7- Mineral extraction (soil and sea).

### 2.5.3 Creative Fermentation Technology for the Future

Fermentation technology is one of the oldest food technology applications that has been developed and utilized for survival. The origin of Asiaticfermentation technology evolved as early as the littoral foragers period of the Primitive Pottery Age (8,000-3,000 B.C.), which led the Neolithic culture of agriculture in the Northeast Asia. The importance of Primitive Pottery Age for the dietary culture of the region was discussed during the 11th IUFoST World Congress in Seoul. Fermentation technology has developed indigenously all around the world by using natural products from the respective region to produce required food materials, from which the characteristic taste and aroma of each cultural society have been made. Steinkraus classified fermentation technology of the world which is responsible for manâs survival into six groups: alcoholic, lactic acid, leavened breads, meat substitutes, meat flavored sauces/pastes and protein/flavoring agents fermentation.

The meat eating habit of Western culture needed food preservation technology in order to keep their perishable meat and milk edible for longer period. Meat sausage, cheese and acid fermented milk making were important food preservation technology methods until the time when refrigerators were available in homes. On the other hand, people who ate cereals as staple food in the East wanted to have meaty flavored and salty condiments, which make the bland taste cereal foods more palatable. This demand led Asian people to develop soybean sauce and fish sauce fermentation technology.

Consumer perception of food has changed through history. Until the 20th century, the food situation of the world was always problematic. Food shortage could occur any time even in the affluent societies. Therefore, man mostly lived for food, and we may call this earlier period the survival food age. In the 20th century access to food no longer was a problem in most of the affluent societies, but time was short for work and leisure. People wanted to save time cooking and obtaining food. Consequently convenience food became the major item in the food market. Now, we all notice that the 21st century is the 'functional food era.' Health oriented food and nutraceuticals are major concerns of today's consumer.

Fermentation technology has adapted itself to social demands. During the survival food age, fermentation was used mainly for food preservation and condiments production. In the convenience food age, it was used for flavor production and other ingredients for industrial mass production of food. The 21st century is called the era of tailor made goods satisfying personal demands, and together with the health benefit demands, fermentation technology finds new challenges in the market. Many of the traditional fermented foods are receiving new attention for their health promoting or disease preventing/curing effects. Scientific evidences for their physiological functions are accumulating and the technologies enhancing the beneficial effects are developing rapidly by using modern biotechnological and genetic engineering techniques.

## 2.6 LACTIC ACID FERMENTATION FOR PROBIOTICS

Yogurt and sour milk products are well known in Western society as probiotic food. Similar products made with rather inexpensive raw materials arefound in many indigenous fermented foods around the world. Korean Kimchi is one of the examples. It is made from vegetables lightly salted (ca. 3%) and fermented by lactic acid bacteria, mainly Leuconostoc mesenteroides. The lactic acid bacteria grown in kimchi could survive in gastric acid and bile juice reaching to the large intestine. More important is that kimchi contains various functional components either originated from the ingredients or formed during fermentation. The same beneficial effects can be expected from African cereal gruels like ogi and uji, Nigerian gari, and Asian vegetable foods like dhamuoi in Vietnam, dakguadong in Thailand and burong mustala in the Philippines and acid fermented seafoods mixed with cereals. In addition to the probiotic effect, these vegetable products have excellent prebiotic function.

Combining Western milk fermentation technology and Eastern cereal processing skills led to the creation of fusion foods like soybean yogurt and rice yogurt. Risogurt is an example of lactic acid fermented vegetable drink studied in Korea made from the mixture of rice and soybean protein. Many of the vegetable yogurt products are found in markets of USA and Japan.

The antimicrobial activity of some lactic acid bacteria against Helicobacter pylori, which is known to cause stomach cancer, is well applied to acid fermented milk beverages and highly accepted by the consumer in Korea. Anti-diarrhea effect of lactic acid bacteria like Lactobacillus acidophilus is enhanced by protein coating. The freeze dried bacterial cells coated with protein are added to food for preventing diarrhea and other health benefits, and this market is increasing rapidly.

### 2.6.1 Soybean Fermentation for Cancer and Degenerative Disease Prevention

The physiological function and health benefit of soybean is widely recognized today. The blood cholesterol level reduction activity of soybean protein and its hydrolysate and the pseudoestrogenic effect of soybean isoflavones are experimentally proved. In addition to flavor formation, soybean fermentation increases the digestibility and the nutritional value, and also produces health benefit functional compounds.

Soybean fermentation for soybean sauce and paste, Korean chongkukjang and Japanese natto production involve the process of enzymatic hydrolysis ofprotein to make peptides and amino acids. Some peptides in soybean sauce and paste are known to have ACE inhibition, antithrombotic and anti-cancer effects. Fermentation with molds, Aspergillus and Rhizopus, converts glycoside form of isoflavones, daidzin and genistin, to aglycones, which have higher potency of estrogenic effect. Production of mucous polysaccharides and enzyme kinase during bacterial (Bacillus) fermentation of soybean is related to the fibrinolytic and immune modulating activity of chongkukjang and natto. By enforcing these components through molecular breeding of the microorganisms involved and novel fermentation skills and improved downstream processes, numerous health oriented food products are produced and also under trial in East Asia and will be launched to the world market in nearfuture.

## 2.7 THE ADVANTAGE OF MIXED CULTURE FERMENTATION

Industrial yogurt production in the West and soybean sauce production in the East were achieved by purely isolated starter culture techniques. However, many of the traditional fermentation starters in the East are naturally fermented mixed culture system. The consumer in their taste and aroma often prefers the traditional fermented products made by the naturally fermented starter culture. The deep and bountiful aroma of Korean rice wine, chongju, and the sharp, strong flavor of Korean soybean sauce, kanjang, are more preferred by Korean people than those of industrial counter parts. The anti-cancer activity of doenjang made from Korean traditional mixed culture starter meju was reported to be higher than that of the industrial product made from koji. The physiological characteristics of microbial strains isolated from traditional fermentation starters and their interactions for growth and biosynthesis of functional compounds are under investigation. More studies on the interaction of different species of microorganism in a fermentation system are needed. Computer aided analysis of microbial interaction will make this sort of study possible and manageable.

### 2.7.1 Safety Issue of Fermented Products

Since fermentation makes raw food materials edible without cooking, therisks of hazardous microbial contamination always exist in fermented food especially naturally fermented traditional foods. The uneven distribution of salt in lactic acid fermented fish products and contamination of Aspergillus flavus in traditional starter cultures for rice wine and soybean sauce may result in severe food poisoning incidences.

On the other hand, most of the traditional fermentation methods have their own inbuilt safeguard mechanisms. The food infectious and poisoning microorganisms contaminated on the vegetables and other raw materials of kimchi are killed within one week of fermentation period mainly due to the acid formation and bacteriocin production.

Also, large amounts of nitrate and secondary amines in vegetable products are reduced by fermentation. The importance of fermentation technology for the improvement of hygienic situation of needed region, where cold-chain system is not well established, was discussed at the FAO/WHO Workshop on the Assessment of Fermentation as a Household Technology for Improving Food Safety held in December 11-15, 1995, Pretoria, Republic of South Africa.

Fermentation technology has confronted new challenge in the era of functional food with its efficient biosynthesis potential. Research for useful strains from traditional fermented foods continues worldwide and relevant information is accumulating. Exchange of knowledge and skills of fermentation technology of the West and the East will accelerate the technology innovation and new product development. Considering the importance of fermentation technology in the era of functional food, the public sectors and international bodies should pay attention to the relevant R&D efforts of academic institutions and industrial research groups. The R&D support of national andinternational funding organizations should include the following research priorities:

1. Nationwide search for useful microbial strains from indigenous fermented products and characterization of their physiological properties.
2. Establishment of a global network on fermentation technology and microbial strains and starter culture information.
3. Studies on the mixed culture fermentation and the interaction of microorganisms in fermentation system.
4. Evaluation of physiological function and health benefit mechanism of the metabolites obtained during fermentation process.
5. Evaluation of safety of fermented products, the microorganisms involved and the metabolites.
6. Evaluation and dissemination of fermentation technology for the improvement of hygienic condition of developing countries.

## 2.8 ADVANCES OF FERMENTATION TECHNOLOGIES

Fermentation technology has been a long-favored biological process due to its simplicity of reaction, high specificity, low costs, and versatility of application. Modern industries have complemented the basic principle of fermentation with advances in techniques and genetic engineering to extend its application to the production of a range of biochemicals, biomolecules, and biofuels. Fermentation's applications are rapidly breaching the confines of food stuff such as wine, cheese, bread, and beer, into high-value chemicals, pharmaceutical ingredients, and food-related chemicals. Further, rising petrol prices and depleting fossil fuel reserves have made a robust case for low-cost and simple fermentation processes for producing drop-in chemical replacements.

"There has been a spurt in research efforts, both in the corporate and academic sector, driven largely by fermentation applications within the energy and utilities, specialty chemicals, and pharmaceuticals industries," says the analyst of this research. "Energy - or more specifically, biofuels - has evinced considerable interest in applying fermentative techniques to produce cost-competitive bioethanol."

While the uses of fermentation techniques are many in modern industrial processes, there are several challenges to overcome before implementing them in large-scale commercial applications. One of the stand-out challenges is that fermentation is a slow, natural bioprocess, which means that time-scale of operations is often quite long. Companies also have to contend with the low yields and poor process economics in large-scale production. "Although fermentation operating costs are lower than full-scale chemical operations, the costs of raw materials and fermentable sugars remain a challenge for most companies," notes the analyst. "In addition, specially designed enzymes and microbes with specific biosynthetic pathways are also expensive." Therefore, the end products derived will be hard pressed to be cost competitive with other petrochemical-derived alternatives.

To stave off these challenges, companies are increasingly developing robust proprietary microorganisms that can ferment the substrates to produce the desired products. Consequently, companies that are able to provide a solution that combines excellence in microbe technologies with superiority in process technologies will be better equipped to explore new application avenues. Importantly, companies that can take advantage of fermentation's unique bio-capabilities, while still remaining economically feasible, will find themselves operating in a market with expansive possibilities. "Many companies have begun engineering innovative enzymes, microbes, and bacterial strains that can help produce biomolecules in bulk," observes the analyst. "Apart from genetic techniques, some of the recent developments that continue to be explored include process-driven technologies such as solid-state fermentation, consolidated bioprocessing, ultrasonication, syngas fermentation, and dark fermentation." These developments are indicative of the huge potential of fermentation as a low-cost and flexible method to produce a range of end products.

## 2.9 WHAT PROCESSES ARE INVOLVED IN MAKING FERMENTED FOODS?

The outcome of a food or beverage fermentation process depends on the microorganism-led conversion of substrates into metabolites that support the aroma and taste, appearance, preservation, and nutritional properties of the finished product. These characteristics are time-dependent and determined by the microbiota as well as by a range of physicochemical parameters, including temperature, pH, water activity, oxidation–reduction potential and substrate availability. How these intrinsic and extrinsic environmental parameters are ultimately managed can have profound effects on the final properties and characteristics of fermented foods.

# INDEX

# INDEX

**D**



**E**



**F**



**G**

# INDEX

# INDEX

Systems and evolutionary biology approaches are now providing a rational basis for controlling or managing microbial diversity and community structure to achieve different fermentation processes. Although fermented foods have long been studied as model systems to understand microbial ecology, these latest efforts integrate broader ecological and evolutionary principles, including dispersal, selection, drift and diversification. The use of these principles enables the control of fermentation microbiota in food independent of whether the fermentation is initiated with starter cultures, spontaneously or by inoculation from a prior successful fermentation of the same type (that is, back-slopping). The application of systems biology approaches combined with community reconstructions can identify specific microbial interactions that drive community composition determine a genetic basis for particular microorganisms to live in a fermented food environment and recreate the domestication processes that generated the industrial cultures used in fermentations. Ultimately, findings from those studies will help to address product variation and quality issues that occur even when starter cultures are used. They might also lead to the identification of biomarkers to monitor these foods throughout production and to predict nutritional and health-impacting qualities.

## 2.9.1 Fermentation and Food Safety

Fermented foods that contain appreciable levels of fermentation-produced organic acids (>100 mM), combined with low water activity, salt, nitrite and other antimicrobials, have a long record of food safety. Likewise, beverages containing 4% or more alcohol and pH values less than 4.5 are also considered microbiologically safe. Many LAB, whether part of the autochthonous microbiota or added as starter cultures, are known to produce bacteriocins that inhibit undesirable bacteria, including Listeria, Staphylococcus and Clostridium.

Food fermentations can also enhance food safety and nutritional quality by removing toxic or anti-nutritive compounds from the raw ingredients. For example, the removal of toxic compounds is a prominent feature of cereal, legume and tuber fermentations. Bitter cassava, for example, contains cyanogenic glycosides that must be removed by fermentation, soaking or other suitable processes to avoid acute toxicity when consumed. During sourdough fermentations, some LAB facilitate the degradation of phytate, a cereal grain-associated compound that chelates divalent cations and prevents their absorption in the gastrointestinal tract. Reducing phytate results in enhanced calcium, magnesium, iron and zinc bioavailability from these breads. Sourdough fermentation is also hypothesized to reduce the concentration of other immune-reactive proteins, including the amylase-trypsin inhibitor in wheat, and could therefore be better tolerated than conventional breads by individuals with non-coeliac wheat intolerance or irritable bowel syndrome.

## Do Fermented Foods have Food Safety Risks?

For any food product, there are safety concerns associated with live pathogenic microorganisms as well as toxins or metabolic products that can produce harmful effects. With few exceptions, food-fermenting LAB, yeasts and filamentous fungi are non-pathogenic and do not produce toxins or harmful end-products. When properly made from safe and wholesome ingredients, fermented foods are rarely associated with gastroenteritis. Nonetheless, some cheeses and low-acid fermented foods can pose a safety risk if the food is contaminated with Listeria monocytogenes, Salmonella, Clostridium botulinum or other foodborne pathogens. Although not a direct effect on safety, some microorganisms, including species of the Lactobacillaceae as well as Enterococcus and Staphylococcus associated with long-ripened cheeses, sausages and other fermented foods, can carry transmissible antibiotic-resistance genes.

The microbial metabolites of some fermented foods can, under certain circumstances, also present safety risks. Alcohol (for example, wine, beer and liquor) and salt (for example, soy sauce or kimchi) are inherent constituents of some fermented foods and should be consumed in moderation. Histamine, tyramine and other biogenic amines are formed by some LAB via the decarboxylation of amino acids during the fermentation of cheese, meats, vegetables, soybeans and wine. In the absence of host-mediated detoxification systems, these amines can cause mild to more severe effects such as migraines. Several strategies have been adopted to reduce or mitigate biogenic amine formation, including hygiene to minimize the occurrence of microorganisms producing these compounds and using decarboxylase-negative starter cultures.

Mycotoxins are a potential concern for all fermented foods produced with filamentous fungi. However, domestication and careful strain selection have effectively eliminated mycotoxin-producing lineages of Aspergillus and Penicillium from koji, cheese and other fermented foods. Other microbial metabolites, including citrulline and reuterin, are precursors of the toxic compounds ethyl carbamate and acrolein, respectively. Both occur in alcoholic beverages as well as in other fermented foods. However, their risks to human health from the exposure to fermented foods have not been established.

### 2.9.2 Fermented Foods and Human Health

Consumer interest in fermented foods has been driven in large part by their suggested nutritional benefits, and this interest has led to renewed popularity of these foods on nearly every continent. However, except for yoghurt and cultured dairy products, few human clinical studies have been performed to verify their benefits. Yoghurt consumption is associated with reductions in adiposity factors, type-2 diabetes mellitus and cardiovascular disease, among other positive indications108. Although much of this evidence is based on prospective or epidemiological studies, more than 20 RCTs with yoghurt and cultured milk products have been reported for both healthy individuals and patient population groups. Likewise, milk kefir, kimchi, sauerkraut, natto, vinegar and sourdough bread have been investigated in at least one RCT. By contrast, evidence of health promotion for other fermented foods is mostly limited to chemical analyses and animal and cell culture models.

A better understanding of the health benefits of fermented foods will be obtained from harvesting information from existing population-based diet and health databases as well as with new RCTs. These studies should address the health outcomes arising from the intake of differentiated fermented food categories, food types, and individual fermented food products with well-characterized strains and nutrient compositions. Large, placebo-controlled RCTs will need to account for the known limitations of these types of nutrition study, including blinding, sample size, diet control, dietary recall and adequate intervention times, as well as the challenges specific to fermented foods. To prevent the foods from being easily distinguished by study participants, placebo controls might need to be made to provide the same sensory attributes expected for the fermented foods being tested. Retrospective cohort or, preferably, prospective cohort studies that meet the Bradford Hill criteria should be used and efforts should be made to avoid misleading or unwarranted conclusions. It should be noted that additional challenges exist for cohort studies because dietary databases do not often include fermented foods as a category and critical aspects of those foods might not be reported.

### *What is the Mechanistic Basis for the Health Benefits of Fermented Food?*

Knowledge on the specific health-promoting properties of fermented foods provides a foundation to evaluate how those properties vary by food type, strain composition and production methods. Several routes for health promotion by fermented foods are proposed (Fig. 3), including nutritive alteration of raw ingredients and the biosynthesis of bioactive compounds, modification of the human gut microbiota, and development and modification of the immune system.

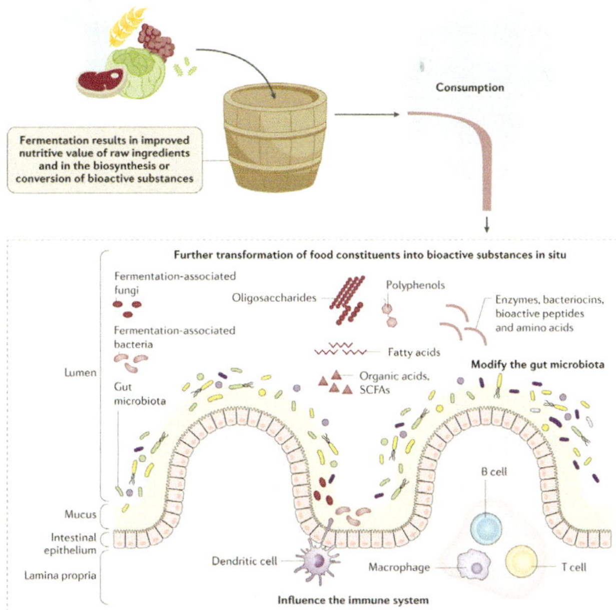


**Fig. 3. Mechanistic basis for the health benefits of fermented foods.**

Health benefits, beyond the nutritional contributions of the raw ingredients, result from the removal, synthesis and transformation of the food components during fermentation by the activities of fermentation-associated microorganisms. Such actions can result in improved nutritive value of the food (for example, through phytate detoxification or vitamin synthesis) or in the generation of biologically active compounds (for example, bioactive peptides or conjugated linoleic acid). Food constituents and fermentation products, along with any remaining viable fermentation microorganisms, are consumed and enter the intestinal tract. Those microorganisms, along with resident members of the gut microbiota, might further transform food constituents in vivo into bioactive substances such as peptides, bacteriocins, amino acids, conjugated linoleic acid or organic acids. The constituents of fermented foods and fermentation-associated microorganisms and their cell products can interact with gut microbiota, the intestinal epithelium or the host immune system. SCFAs, short-chain fatty acids.

Microbial activity during food fermentations results in the enrichment and/or removal of compounds that affect the nutritional composition of the final food product. Microorganisms reduce the concentrations of high-calorie monosaccharides and disaccharides (glucose, sucrose and fructose) present in milk, meat and plants via catabolic pathways. Reductions in certain sugars could also reduce the glycaemic index and improve food tolerability (for example, lactose in dairy foods, fructans in wheat, or raffinose, stachyose and verbascose in soybeans and legumes). Fermentation can result in the hydrolysis of polysaccharides, proteins or fats, thereby increasing their digestion. Other enzymatic transformations with important nutritional implications also occur, including detoxification reactions (for example, degradation of linamarin in bitter cassava) and the removal of anti-nutritive factors (for example, inactivation of trypsin inhibitor in soybeans and phytic acid in cereals such as sorghum). For polyphenol-containing foods, the conversion of phenolic compounds by lactobacilli1 increases the bioavailability of flavonoids, tannins and other bioactive compounds. The biosynthesis of vitamins, amino acid derivatives, organic acids and cofactors can also occur during fermentation, with effects at either local gastrointestin al or systemic sites. Some of these compounds are broadly distributed between fermented food types (such as lactic acid and acetic acid), whereas others are common in certain foods (for example, alkyl catechols) or limited to certain microorganisms with specific enzymatic activities (for example, synthesis of γ-aminobutyric acid, conjugated linoleic acid or angiotensin-converting enzyme inhibitors).

Multiple studies in humans have shown that microorganisms in fermented foods can survive gastric transit and reach the colon. Indeed, many of the LAB that dominate lactic acid-fermented foods possess intrinsic characteristics that promote their ability to survive gastric transit (for example, acid and bile tolerance). Depending on individual dietary habits, fermented food-associated LAB can transiently constitute between 0.1% and 1% of the bacteria in the large intestine and a comparable proportion in the small intestine. This percentage is based on current estimates of autochthonous microbiota in the gastrointestinal tract146 and the presence of up to 1011 LAB cells in a single serving of many fermented foods, such as yoghurt or kefir, that contain live and active microorganisms. Similarly, another study published in 2020 showed that food-associated LAB reached faecal metagenome abundances of >0.1%. Although these microorganisms are unlikely to maintain long-term residence in the intestine, some fermented food microorganisms are known to be metabolically active in the gastrointestinal tract and short-term colonization could be sufficient to synthesize bioactive compounds, inhibit intestinal pathogens and mediate epithelial modulatory effects (for example, via interaction with Toll-like receptors). Such interactions would be augmented by the repeated daily consumption of the fermented food. According to population-based studies and RCTs, fermented foods can also influence the composition of the gut microbiota. Modulation of the gut microbiota can result from the living (or inactivated) microorganisms in those foods, the nutritional components and metabolites released as a result of fermentation, and changes these food constituents confer on the host immune system. These effects are likely dependent on inter-individual differences in host physiology and gut microbiota composition.

As approximately 70% of the human immune system is located in the gastrointestinal tract, foods and beverages are the major conduit of contact between external antigens and the human body. The gastrointestinal tract is vulnerable to the initial pattern of microbial colonization during the first months of life, potentially setting a critical window for microbial stimuli effects on the immune system. In one cross-sectional study, fermented food intake (fermented vegetables) during early childhood was associated with a reduced risk of childhood atopy. In another epidemiological study, fermented food consumption combined with common daily-life exposure (for example, hand versus machine dishwashing) also reduced the risk of childhood allergies. The authors from the former study further reported that an anthroposophic lifestyle was associated with differences in infant microbiome structure, including a higher abundance and diversity of LAB, and a higher concentration of acetate compared with infants from a traditional lifestyle. Fermented food intake is also one of the synergistic factors associated with a farming upbringing, a lifestyle factor that has consistently been associated with reduced allergy and asthma risk. These associations could indicate that a lack of fermented foods in modern, industrialized societies constitutes a substantial loss in exposure to non-harmful microorganisms important for immune system development and maintenance.

Although fermented foods such as milk kefir have been shown to modulate immune responses in numerous animal models, RCTs or prospective studies on the human immune system have yet to be performed. It is expected that the modulation of the human immune system by fermented foods would be the result of the combined effects of compounds present in the starting ingredients and those formed during fermentation as well as of living and dead or inactivated microorganisms. Those fermentation-associated microorganisms and their cell components (for example, peptidoglycan, surface proteins, exopolysaccharides and lipoteichoic acid) are already known to be immune reactive according to animal model and in vitro studies. Knowledge about other immune-modulating compounds, such as D-phenyllactic acid, produced by lactic acid bacteria in situ, is still emerging. Ultimately, the precise molecular stimuli in fermented foods responsible for immunomodulation probably depend on the total composition of the producrt.

### *What are the Regulatory Considerations for Fermented Foods?*

Guidelines that govern food fermentation are covered in international regulations and are mainly concerned with food safety. The use of microbial cultures is also regulated and includes criteria for establishing safety, such as the 'Generally Recognized As Safe' designation in the USA or the 'Qualified Presumption of Safety' list in Europe. The latter, for example, is a designation assigned by the European Food and Safety Authority to groups of microorganisms that, in general, do not raise safety concerns as components of foods, including fermented foods. Strains developed by the use of recombinant DNA technology or those that are genetically modified have different regulatory controls. For example, in the USA, genetically modified strains must have a 'Generally Recognized As Safe' status, whereas in Europe, such products require Qualified Presumption of Safety status.

The identification of core microbial components in fermented foods has the potential to lead to new regulations around the labelling of these foods. Regulations could be used to ensure that minimum requirements relating to the involvement of specific microbial taxa in the fermentation process are met. Only a few standards exist, mostly for cultured dairy products. For example, the Codex Alimentarius states that yoghurt should be made using a combination of S. thermophilus and L. delbrueckii subsp. bulgaricus and that kefir is a fermented milk consisting of Lentilactobacillus kefiri and species of the genera Leuconostoc, Lactococcus and Acetobacter, in addition to lactose-fermenting yeasts and non-lactose-fermenting yeasts. Similar standards could emerge as the microorganisms present in other fermented foods are identified.

### *What is the Standing of Fermented Foods in Dietary Guidelines?*

Fermented foods are widely consumed around the world and have been estimated to account for approximately one-third of the human diet. However, with few exceptions, fermented foods are generally absent as a recommended category in dietary guidelines. The only country, to our knowledge, that has a specific guideline is India, which encourages pregnant women to consume fermented foods. Other countries, including the USA and Canada, but there is no specific emphasis on fermented foods. Owing to the high levels of live, potentially health-promoting microorganisms in many fermented foods, these foods have been advocated for inclusion in dietary recommendations. To advance this field, studies that collect dietary information should also track foods that contain live cultures. Adding granularity to dietary intake data so that fermented foods are not subsumed under other categories will enable researchers to better understand the role of these foods in health.

# 3

# BIO-FERMENTATION TECHNOLOGY

## 3.1 INTRODUCTION

Biotechnological processes are conducted in a bioreactor, where substances are being synthesized with the help of bacterial, yeast- and cell strains. The end product is used in many areas of our daily lives, like yoghurt and beer brewing processes. Besides in food and beverage applications, it is also used in the field of medical drug research and production. Flow controllers play an important role in these Bioreactors. The rise of biotechnology and its growing field of applications are enough reason to find out more about this fascinating technology.

Fermentation is the breaking down of sugar molecules into simpler compounds to produce substances that can be used in making chemical energy. Chemical energy, typically in the form of ATP, is important as it drives various biological processes. Fermentation does not use oxygen; thus, it is "anaerobic".

The term biotechnology came into general use in the mid 1970s, gradually superseding the more ambiguous `bioengineering', which was variously used, to describe chemical engineering processes using organisms and/or their products, particularly fermenter design, control, product recovery and purification. Most scientists agree that all processes that utilize biological organisms constitute biotechnology, but what is disputed is which processes do not. Definitions of technology too vary from the simple 'applied science' to 'the scientific study of the practical or industrial arts'.

In consequence of the disagreements, there are at least ten different definitions of biotechnology, each qualified according to the context of use. The following are among the more widely used definitions of biotechnology:

(a) "The application of biological organisms, systems or processes to manufacturing and service industry".

(b) "The application of scientific and engineering principles to the processing of materials by biological agents to provide goods and services". Agents include a wide range of biological substances, such as enzymes, as well as whole cells or multicellular organisms. 'Goods and services' covers processes such as waste and water treatment.

(c) "The integrated use of biochemistry, microbiology and chemical engineering to exploit plant materials and genetic resources for the production of specific products and services".

Many definitions are rather vague about the nature of the organism or agents involved in biotechnology. It is argued that systematic farming of plants and animals for food and fuel also falls within these definitions.

Ancient fermented food processes, such as making bread, wine, cheese, curds, idli, dosa, etc., some of which are some 6,000 yr old, and developed long before man had any knowledge of the existence of the micro-organisms involved, also genuinely constitute biotechnology. However, for the sake of convenience, many people exclude these traditional processes from the realm of biotechnology. Conventional agriculture is a well-developed industry in its own right, but in practice, this is not included in biotechnology. Aspects of `modern biotechnology' may have significant effects on 'traditional biotechnology'. Genetic manipulation to improve brewing and baking yeasts or to introduce new characteristics in crops, biological control of plant pests, and new methods of diagnosing and preventing plant, human and animal disease, are all now realisable. Whatever the definition, experimental production of new varieties of organisms, is one of the important objectives of biotechnology.

In simpler words, biotechnology means the industry-scale use of organisms and/or their products. Now biotechnology virtually includes the scientific, technological and commercial aspects of almost every area of human welfare from agricultural production to pollution control.

Bio-fermentation technology makes it possible to grow a number of fungi in large tanks and, in a matter of days, a large quantity can be produced. The technology has now become highly advanced and is making previously rare herbs like Cordyceps and Ganoderma much more accessible. Many studies indicate that the chemical nature of this biotechnology Cordyceps is almost identical to that of the wild variety. The Cordyceps contained in most commercial products is produced by this technology.

Fermentation term for the first time was coined by Louis Pasteur for a phenomenon of bubbling of sugar solution. Later on, it has been applied for the phenomenon of production of different chemicals involving microorganisms.

Presently, the term is used solely to any phenomenon involving microorganisms. Many products are made by large-scale fermentation including amino acids, enzymes, organic acids, vitamins, antibiotics, solvents and fuels. The typical fermentation process is depicted in Fig. 1.

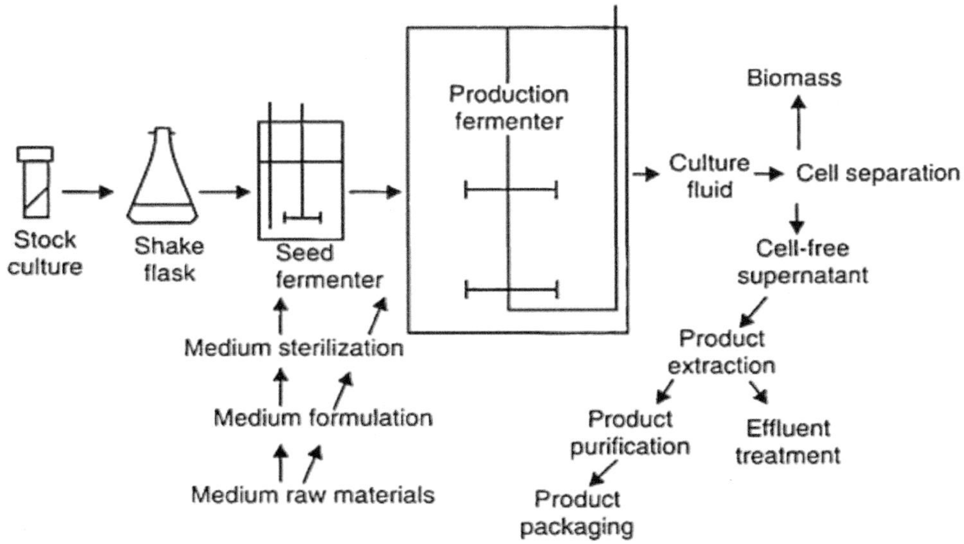


**Fig. 1. A Schematic Repressentation of a Typical Fementation Process**

The advantages in producing materials by fermentation are as follows:

1. Complex molecules such as antibiotics, enzymes and vitamins are impossible to produce chemically.
2. Optically active compounds such as amino acids and organic acids are difficult to prepare chemically.
3. Though some of the products that can be economically derived by chemical processes, but for food purpose they are better produced by fermentation such as beverages, ethanol and vinegar (acetic acid).
4. Fermentation usually uses renewable feed stocks instead of petrochemicals.
5. Reaction conditions are mild, in aqueous media and most reaction steps occur in one vessel.
6. Byproducts of fermentation are usually chemicals. The cell mass and other major by products are highly nutritious and can be used in animal feeds.

However, it is beset with some drawbacks, which are as follows:

1. The products are made in complex solutions in low concentrations as compared to chemically derived compounds.
2. It is difficult and expensive to purify the product.
3. Microbial processes are much slower than chemical processes, increasing the fixed cost of the process.

4. Microbial processes, are subjected to contamination by competing microorganisms, requires the sterilization of the raw materials and the containment of the process to avoid contamination.
5. Most microorganisms do not tolerate wide variation in temperature, pH and are also sensitive to upsets in the oxygen and nutrient levels. Such upsets not only slow the process, but fatal to microorganism. Thus careful control of pH, nutrients, air and agitation require close monitoring and control.
6. Although nontoxic, waste products have high BOD and requires extensive sewage treatment.

Though microorganism belonging to bacteria, fungi and yeasts are extensively used in these fermentation, few fermentations are also based on algae, plants and animal cells.

Several cellular activities contribute to fermentation products such as:

1. Primary metabolites Ethanol, lactic acid and acetic acid.
2. Energy storage compounds Glycerol, polymers and polysaccharides.
3. Proteins SCP, enzymes of both extra and intracellular nature and foreign protein.
4. Intermediate metabolites Amino acids, citric acid, vitamins and malic acid.
5. Secondary metabolites Antibiotics\
6. Whole cell products SCP, bakers yeast, brewers yeast, bio-insecticides.

Some of the products such as ethanol, lactic acid and cell mass products are generally growth associated, while secondary metabolites, energy storage compounds, and polymers are non-growth associated. Other products, such as protein depends on the cellular or metabolic function. Unlike primary metabolites which are essential for growth and reproduction, secondary metabolites are not essential for the growth and development of reproducing organism and are produced only in luxuriant conditions.

The secondary metabolites basically are:

1. Secondary metabolites are produced only by few organisms.
2. Secondary metabolites are needed depending on environmental conditions.
3. Secondary metabolites are produced as a group of closely related structures.
4. Some organisms forms a variety of different classes of substances such as secondary metabolites.
5. The regulation of biosynthesis of secondary metabolites differs significantly from that of primary metabolites.
6. Secondary metabolites are mostly produced in iodophase (Fig. 2)

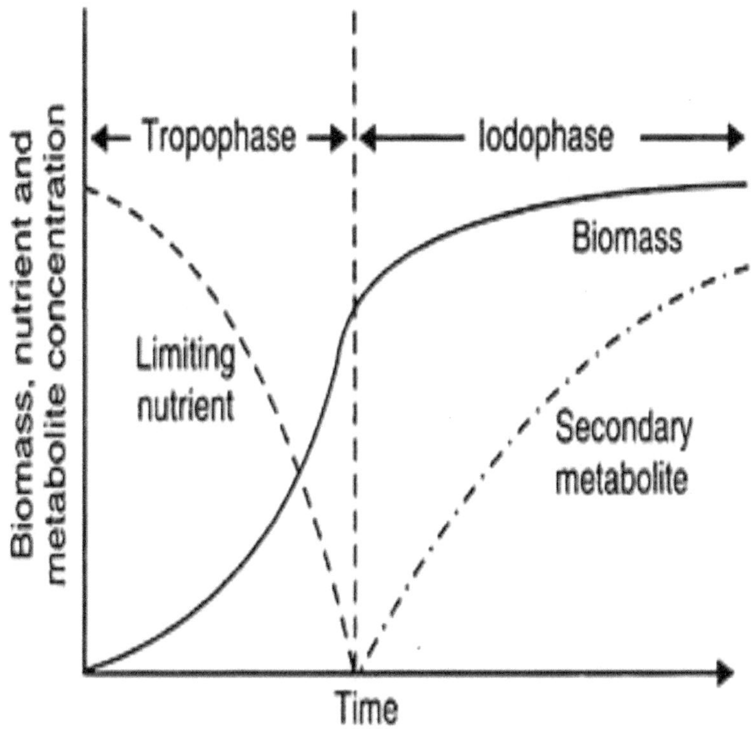


**Fig. 2. The growth phasses of biomass production and secondary metabolite production**

Origin and production of different secondary metabolites are depicted in Fig. 2. and 4.

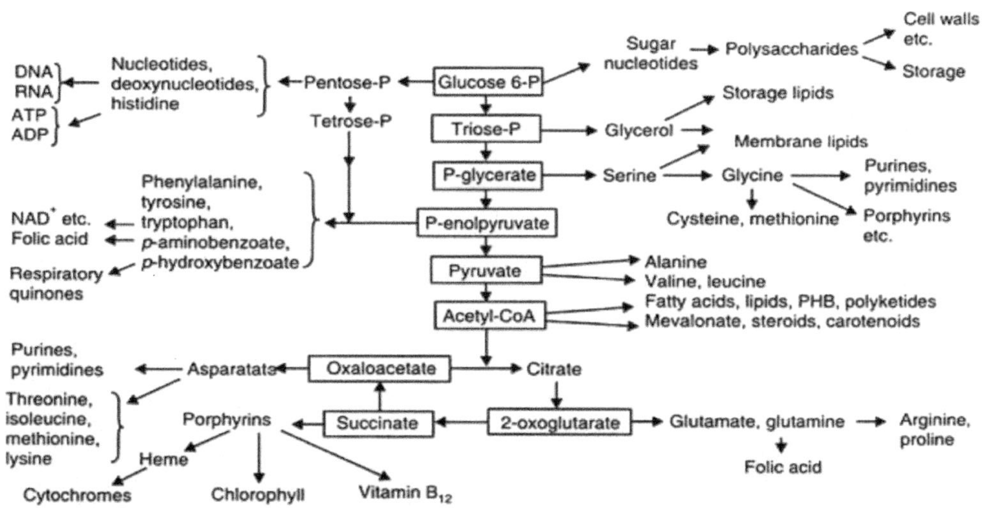


**Fig. 3. Primary metabolites giving rise to variety of cell substances**

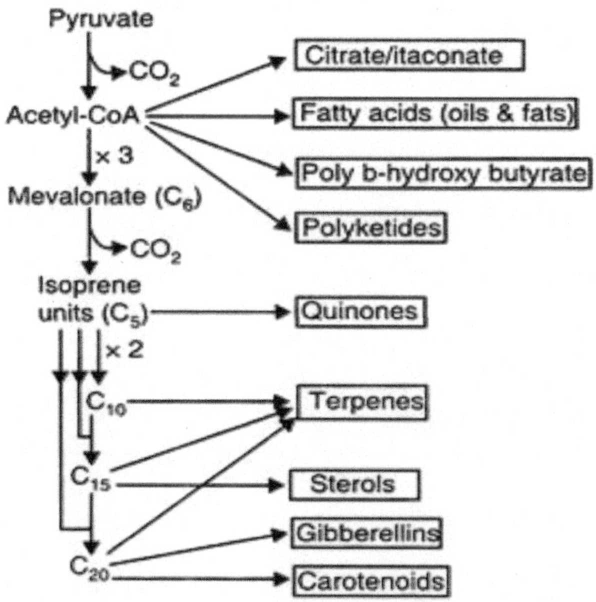


**Fig. 4. Production secondary metabolites**

Fermentative products are in use by man since ancient times. Fermentation of grains or fruit produce, bread, beer and wine that retained much of the nutrition of raw materials, while keeping the product from spoiling. The natural yeasts that caused fermentation added some vitamins and other nutrients to the bread or beverage.

Lactic acid producing bacteria ferment milk to yogurt and cheese and extend the life of milk products. Other food products such as pickles, vegetables and the fermentation of tea leaves and coffee beans were preserved or enhanced in flavor by fermentation.

## 3.2 HISTORICAL EVENTS IN THE PROGRESS OF FERMENTATION

Fermentation was an art until the second half of the 19th century. A batch was begun with either a starter, a small portion of previous culture, or with culture residing in the products or vessel. Pasteur made it clear that fermentation needs, heat treatment to improve storage quality and thus formed the basis for sterilization of medium. Emil Christian Hansen used for the first time pure culture of yeast for production of yeast in Denmark.

During 1920-30 the emphasis in fermentation shifted to organic acids primarily lactic acid and citric acid. The discovery of penicillin in 1929 and commercialized in 1942, gave a boost to fermentation industry and led to the development of big fermenters and submerged cultivation. Success of penicillin inspired pharmaceutical companies to launch massive efforts to discover and develop many other antibiotics.

In 1960s amino acid fermentations were developed in Japan. Commercial production of enzymes for use in industrial process began on a large scale in 1970. The discovery of the tools of genetic engineering expanded the possibilities for products made by fermentation in situ, and the first genetically engineered fermentation product was developed and commercialized in 1977. The historical events developed in the progress of fermentations are precised in Table 1.

**Table 1: Historical Events in the Progress of Fermentarion**

| Year of commer-cialization | Fine Chemicals | Enzymes | Therapeutic substances | Microbial transformations | Other |
|---|---|---|---|---|---|
| 1880-1900 | Lactic acid | - | - | - | - |
| 1900-1910 | Ethanol Glycerol | Amylases | - | - | - |
| 1910-1920 | Acetone n-butanol | Invertase | - | - | - |
| 1920-1930 | Citric acid | - | - | - | - |
| 1930-1940 | Gluconic acid | Proteinases | Riboflavin | Sorbose Acyloions | - |
| 1940-1950 | Itaconic acid 2-keto-D-gluconic acid | Cellulases Pectinases | Vitamin $B_{12}$ Penicillin G Other antibiotics | - | - |
| 1950-1960 | Kojic acid L-glutamic acid lysine | Oxidase Glucose Catalase | Ampholericin B Erythromycin Semi-synthetic penicillins | Steroid oxidation | Gibberellins Dextran SCP |
| 1960-1970 | Valine | Glucose Isomerase Glucamylase Lipases Lactase | Amphomycin Cephalosporins Fusidic acid Linomycin Some more antibiotics | Acetone dihydroxy | Xanthan 5'nucleosides bioinsecticide |
| 1970-1980 | Comenic acid | Microbial enzyme Rennet Replacement Melibiase Dextranase | Salinomycin Validanycin A | Sterol cleavage Xylitol, malic acid | Ribose zearalenol |

Fermentation may be aerobic if it is operated in the presence of oxygen, while it may be anaerobic if carried out in the absence of oxygen. Anaerobic fermentations can be carried out either by use of fresh medium, covered with an inert gas such as nitrogen or argon or accumulation of $CO_2$ or foam (Fig. 5).

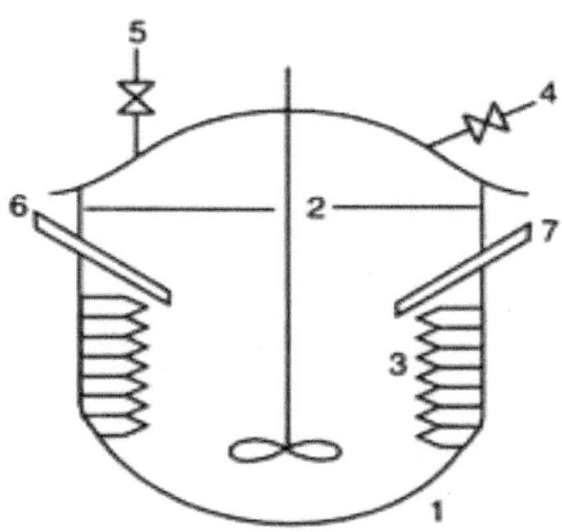


**Fig. 5. Anaerobic Fermenter**

The fermentation is called batch fermentation when it is operated for a definite period. On the other hand, fermentation which is operated for an indefinite period it is called continuous fermentation. Some of the organisms are sensitive to substrate concentration and they are inhibited when the substratum is in high concentration. Under such conditions, fermentation can be carried by addition of substrates in installments and the process is called Fed batch fermentation.

Fermentations can be carried out under non-aseptic conditions where the risk of contamination is not a major concern. However, fermenters must be designed for prolonged aseptic operation. The design rules for an aseptic bioreactor demand that there is no direct contact between the sterile and non-sterile sections to eliminate microbial contamination.

Similarly, fermentation based on number of organisms involved can be classified into simple fermentation when only one organism is involved to produce a product from substratum. On the other hand, in some fermentations two organisms are involved in order to get a fermentation product from a substratum. The product of first phase of fermentation serves as substratum for second phase in order to yield desired product.

In this type of fermentation, two organisms may grow simultaneously and product is formed instantly. Commercial growth of lichens involving algae and fungi is a good example for simultaneous fermentation. Production of glutamic acid from glucose firstly gets oxidized to ketoglutaric acid, which in turn get aminated to produce glutamic acid and production of lactic acid from glucose by yeast and Lactobacillus lactis, production of β-carotene jointly by (+) and (-) strains of either choaenophoracucurbitarum or Blakesleea trispora are three very good examples.

On the other hand, the two organisms involved in a fermentation are separated widely in time and space, such fermentation is called successive fermentation. For example-production of acetic acid from glucose. First glucose is acted by yeast to produce ethyl alcohol, which is oxidized to acetic acid by Acetobacter aceti. Similarly production of lysine from glycerol. Glycerol is fermented to Diaminopimelic acid (DAPA) by an auxotrophic mutant of E. coli which gets aminated to form L-Lysine by Aerobacter aerogenes. When more than two organisms are involved in a fermentation it is called as mixed fermentation or multiple fermentation.

In this fermentation, the substratum is heterogeneous and organisms with different potentialities of producing enzymes are involved in the fermentation. For instance, degradation of municipal wastes and decomposition of dead plants and animals can be taken as mixed or multiple fermentation. Similarly, remediation of waste water comes under this fermentation.

## 3.3 PROCESS OF FERMENTATION

Fermentation process can be conveniently divided into six stages regardless of the type of process.

They are:

1. The formulation media used for the growth of the microorganism to be employed as inoculum and also in the production of fermentation products.
2. The sterilization of the medium, fermenter and other associated equipment.
3. The preparation of adequate quantities of pure culture that is to be inoculated into the fermenter.
4. The creation of optimum conditions in the fermenter for optimum growth of the organism and for optimum output of the desired product.
5. The extraction of the product and its purification.
6. The disposal of effluents generated during fermentation.

The inter relationships among these six phases are diagrammatically illustrated in Fig. 2.10.

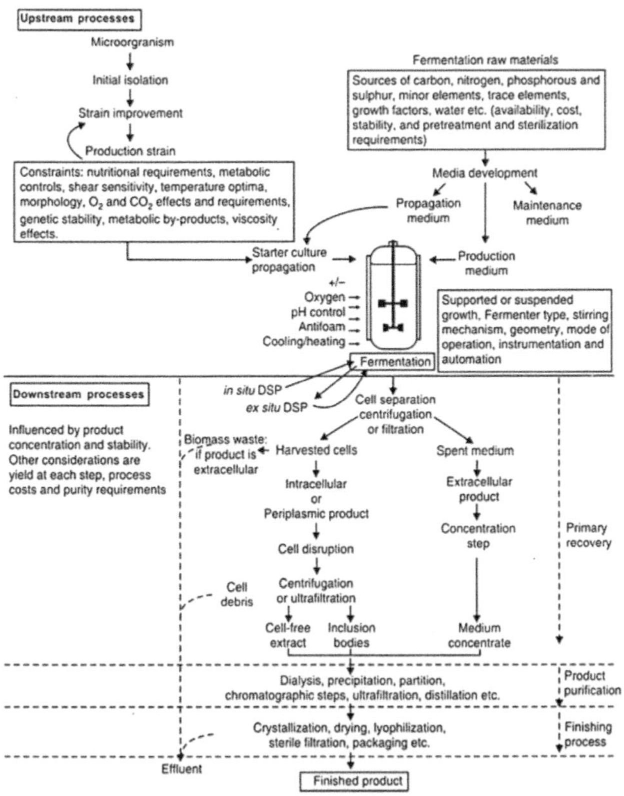


**Fig. 6. The upstream and downstream process of a typical fermentation process.**

This process varies with the type of organism used and product to be produced.

**The entire process can be discussed under two headings:**

**(a) Upstream process:** It includes selection of organism and medium, medium sterilization, inoculation and ends with monitoring of fermentation process and product formation. This involves selection of microorganism. The selection of microorganisms for fermentation should be critically done. At first it should have potential to produce particular substance in an economic amounts. It should be nonpathogenic and non-hazardous. Further it should be amenable to growth in a fermenter and produce the product in good amounts,

**(b) Downstream process:** It includes the product separation and purification and effluent treatment.

## 3. 4 ADVANTAGES OF BIO-FERMENTATION

### 3.4.1 Standardization

Highly controlled by scientific means during its growth and "standardized", every batch is virtually identical and, in a sense, "perfect." This is very important in pharmaceutical terms because without standardization, it is difficult or impossible to develop drug-type standards for substances like Cordyceps. Once a herb can be standardized, all kinds of studies can be conducted that will be accepted by the scientific community. However, standardization, by itself, does not make a product effective.

### 3.4.2 Cost

Products can be produced more cheaply by bio-fermentation. High quality wild Cordyceps is rare and now sells for well over a thousand dollars a pound in cities like Hong Kong, Tokyo, New York or London. The culture-grown Cordyceps is available at a third of the cost, with approximately the same benefits. Fermentation technology makes this substance more readily available.

### 3.4.3 Vegetarian

Wild Cordyceps, by weight, is mostly caterpillar. For vegetarians, and those who do not want to consume caterpillars, the new technology provides a solution. The fermentation technology does not include the caterpillar in the growing process. The fungus is grown without the use of animal nutrients and the result is a 100% pure "vegetarian" health product.

### 3.4.4 Wild Cordyceps

Studies show that wild Cordyceps is richer in certain components and the proportion of components is different, which probably makes a difference in activity. Cultivated Cordyceps is still a premium tonic herb and the Cordyceps most commonly found in Chinese herb stores is cultivated.

### 3.4.5 Supply of Bio-Fermentation Technology

Fermentation technology to make crop stalks and straw waste into a green pollution-free bio-feed for livestock breeding, animal manure and other waste but also rural and urban solid waste by biological agents quickly look into a high bio-organic fertilizer into the green plant. Agent can look to a variety of fermentation of crop straw, melon vine, animal manure, leaves, weeds, and bran residue of vinegar, wine dregs, residue of vinegar, soy sauce residue, soybean, mushroom residue, powder residue, rubbish, bone meal, bagasse and other waste into organic fertilizer and bio-fast feed.

Fermentation of organic fertilizer and a ton of finished feed, just use 2 kg of bio complex microbial agents, at least a week, as many as two weeks, all green waste can be turned into pollution-free eco-fertilizers and bio feed. The fertilizer and feed production is simple and low cost. Use of their houses in rural areas, idle houses and fields, can be produced throughout the year to reduce the human transport, local production, local use. Is the majority of farmers do not buy fertilizer and feed friends, turning waste into the wealth channel.

Fertilizer complex microbial agents 18 yuan per kg, straw fermentation agents 30 yuan per kg, 25 kg/bag from the mail, free shipping over 1 ton and 10 ton, the production company free of charge technical services. Look biotechnology is organic fertilizer plant, feed mill, factory farming, sewage treatment plant, township, town, village, ideal for entrepreneurial farmers.

## 3.5 INDUSTRIAL PRODUCTION OF ETHANOL

### 3.5.1 History of Ethanol Production

- Louis Pasteur demonstrated the fermentation of sugar by microbes and their regulations as well. He found that the yeast cells produce 20 times more cell materials under aerobic conditions than under unaerobic conditions. He explained that Oxygen inhibits fermentation process. Yeasts are aerobic microorganism, but glucose fermentation takes place under unaerobic condition. This is also known as "Pasteur effect".
- In 1815, Gay Lussac formulated the conversion of glucose to ethanol.
- During 1980, lactic acid was produced by fermentation as a substitute for tartaric acid used in baking powder which was not commercialized.

### 3.5.2 Characteristics of Ethanol

- Ethanol is a volatile, flammable, colourless liquid.
- Ethanol is a straight-chain alcohol, and its molecular formula is $C_2H_5OH$.
- Its empirical formula is $C_2H_6O$, making it a constitutional isomer of dimethyl ether.
- The fermentation of sugar into ethanol is one of the earliest organic reactions employed by humanity

### 3.5.3 Production

The entire process of ethanol production is divided into two stages namely upstream process and downstream process.

### 3.5.4 Upstream Processing

Upstream processing encompasses any technology that leads to the synthesis of a product as well as the fundamental science and engineering needed to understand product formation. Upstream includes the exploration, development and production of crude oil, bitumen and natural gas etc.

### 3.5.5 Types of Ethanol Production

Ethanol is produced both as a petrochemical, through the hydration of ethylene, and biologically, by fermenting sugars with yeast.

### 3.5.6 Ethanol Hydration

Ethanol hydration is most often made from petrochemical feed stocks, typically by the acid-catalyzed hydration of ethylene. The catalyst is most commonly phosphoric acid absorbed on to a porous support such as diatomaceous earth or charcoal. This catalyst was first used for large scale production by shell Oil Company in 1947. The reaction is carried out with an excess of high pressure steam at 300°C.

### 3.5.7 Microorganisms Used in Ethanol Production

Some of the microorganisms that are used in the fermentation process are mentioned below:

Bacteria: Clostridium acetobutylicum; Klebsiella pneumoniae; Leuconostoc mesenteroides etc.

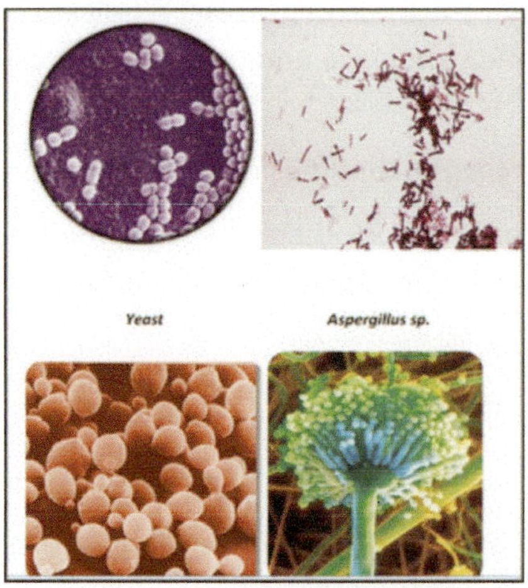


**Fig. 7. Aspergillus oryza; Mucor sp; Neurospora crassa; Saccharomyces sp. (yeast); etc.**

### 3.5.8 Fermentation Conditions

**Carbon Sources:** Sugar concentration in the range of 10 to 18 % is found satisfactory. Sugar concentration of 12 % is often used. For example, cane molasses contain sucrose as a carbon source. High concentration affects yeast adversely while low concentration makes the process uneconomical.

**Nitrogen Sources:** Many organic as well as inorganic nitrogenous compounds may be used. Ammonium sulphate is generally used. The level of nitrogen in the medium is limited.

**Growth Factors:** It is not necessary to add the growth factors separately in the preparation of the fermentation medium. Since the raw materials used satisfy the requirements of the culture.

**PH:** PH of the fermentation medium is 4.8 to 5. The higher PH increases the chances of contamination, whereas lower PH values inhibit the yield of ethyl alcohol.

**Temperature:** Usually, the temperature range of 70 to 80°F is preferred. Temperature can be controlled by cold water spray around the fermentation tank. Higher temperature favours the growth of bacteria and also causes loss of alcohol due to evaporation.

**Fermentaion Method:** Fermentation of ethanol is carried out in large fermentation tanks. The Inoculum of microorganisms is maintained in the fermenter at the optimum growth conditions. At the same time fermentation medium is formulated and transferred to the sterile fermenter. Liquid medium in fermenter is inoculated with a small amount of inoculum of yeast. Fermentation starts within a few hours after the addition of yeast. The process becomes rapid after 24 hours to complete. Duration of ethyl alcohol fermentation is usually 30 to 72 hours when the specific gravity of the fermented liquid becomes constant.

## 3.6 BIO-FERMENTATION TECHNOLOGY TO IMPROVE EFFICIENCY OF SWINE NUTRITION

In ancient times, mankind survived with food from daily hunting. As human being accumulates experiences and knowledge about nature, agriculture became a major way of obtaining food. Agricultural activity also allowed us to have stable food supply and this was one of important reasons for the increase in human population and cultural flourish. As entering modern society, growth of human population accelerated as so demands for food. Limited space and resources with fast growing demands for food has pushed agriculturists to improve efficiency of food production through agriculture. Animal agriculture (or livestock production) is an important part of agriculture producing meat and fat. Pork production takes a key role providing over 42% of global meat consumption.

### 3.6.1 Swine Nutrtion and Biotechnology

We can't ignore that continuous genetic selection has made great contribution to animal agriculture by improving genetic potentials of maximal productivity. Scientific advances in nutrition, then, allowed expressing these genetic potentials to maximize productivity of livestock. It is amazing to see the improvement of productivity of pork production during the 20-40 years. The author had a chance to compare few economically important traits of pigs (especially for sows) between 1979 and 1999.

During the past 20 years, performance of sows has been improved dramatically. As examples, litter size of sows has increased by 9% whereas litter mortality has reduced by 18%.

It is not difficult to see the evidences showing that biotechnology was a key driving factor for the great improvement in the efficiency of pork production. One can easily think genetic engineering as a main theme of biotechnology. However, genetic engineering is only a portion of biotechnology. United Nations Convention on Biological Diversity defines biotechnology as "Any technological application that uses biological systems, dead organisms, or derivatives thereof, to make or modify products or processes for specific use". Biotechnology indeed includes a broad range of applications including bioengineering, bio-remediation, gene therapy, genetic selection, bio-degradation, cloning, bio-fermentation, etc.

Among these, we can find many evidences that pork production has much been benefited from application of bio-fermentation technology in preparing feed and feed additives. In the next sections, few examples of successful application of bio-fermentation technology in swine nutrition are described based on the author's own research experiences.

## 3.7 SUPPLEMENTAL AMINO ACIDS

It is early 1960s when an amino acid, mono-sodium glutamate, was produced in a large scale using biofermentation technology to be used for food application. Soon after, similar bio-fermentation technology was applied to produce feed grade supplemental amino acids.

Supplemental amino acids are produced by liquid state fermentation. Briefly, sugars from sugar cane are extracted and used as a substrate for fermentation by microorganisms specifically selected for synthesizing targeting amino acids. After these microorganisms excrete targeting amino acids to broth, drying process and filtration are followed to obtain crystalline amino acids. Good examples of supplemental amino acids that can be produced in large scales are Llysine HCl, L-threonine, and L-tryptophan which are, in fact, three key limiting essential amino acids for swine. In recent years, feed grade supplemental L-glutamine and Lvaline are also introduced by a company leading this area. L-arginine HCl is another good example of supplemental amino acid for its unique function in an animal body that is also produced by bio-fermentation.

Production efficiency of supplemental amino acids through bio-fermentation has been improved significantly due to advances in biotechnology. Currently production efficiency of L-lysine reaches over 50%.

L-lysine HCl is the first feed grade supplemental amino acid that is produced in a large scale by bio-fermentation which is soon after production of mono-sodium glutamate for food application. It was early 1960s when feed grade Llysine HCl was mass-produced first time in Japan. Since then L-lysine has well been used as the first limiting essential amino acid in swine feed. It may be 1950 when the first research paper about the requirement of lysine for swine published in a scientific journal.

Production efficiency of lean pork has been maximized as nutritionists can formulate swine feed matching requirements of these limiting essential amino acids and reducing crude protein contents. It can be formulated to include 11.7% crude protein with the use of L-lysine, Lthreonine, and L-tryptophan which would provide an equal nutritional value of feed containing 15.2% crude protein without supplemental amino acids.

Recently, a large scale production of feed grade L-valine has been initiated using bio-fermentation technology. Use of L-valine, which is often 4th or 5th limiting essential amino acids for swine, allows nutritionists to further reduce crude protein content in the feed without affecting growth performance of pigs. When these four feed grade amino acids are used in swine feed, 1% value of crude protein content can be further reduced. Currently feed grade Lvaline is only marketed in European Union, however, the use of L-valine will soon be extended to Americas and Asian countries. Development of new feed grade L-valine allowed nutritionists to formulate swine feed with high accuracy for maximal lean growth as well as that is environmentally friendly by reducing contents of excess amino acids in the diets over the requirements.

Another feed grade amino acid that is recently development by bio-fermentation is L-glutamine. Concept of applying L-glutamine in swine feed is somewhat different from the use of previous four limiting amino acids.

L-glutamine is known as a non-essential amino acid that is not a limiting amino acid for swine. However, recent research clearly indicates specific function of L-glutamine other than being used for protein synthesis in an animal body. It is demonstrated that L-glutamine is a major energy substrate and an essential precursor of Nacetylglucosamine-6-phosphate for proliferation and growth of enterocytes and for prevention of intestinal atrophy. L-glutamine is essential for development of gut function especially for newly weaned young piglets that are under stressful condition due to new environment and diet changes. It is also suggested that pigs under chronic or acute stress with immune depression can be benefitted by supplemental L-glutamine as immune cells use L-glutamine as a major energy source.

L-arginine is another important example of an amino acid that has great potential to be used in swine feed even though a large scale feed grade production has not occurred yet. There are many research papers showing unique functions of L-arginine in an animal body including humans.

Briefly, L-arginine can be used by endothelial cells lining blood vessels which possess an enzyme called nitric oxide synthase converting L-arginine to citrulline releasing nitric oxide. Production of nitric oxide causes conversion of cellular GTP to cGMP which will cause vasodilatation with increase blood flow. In the mean time, L-arginine can also be converted to orinithine which will in turn metabolized to a group of polyamines including spermine and spermidine.

Polyamine is known as a key components used for angiogenesis and cell proliferation. A recent study showed that dietary supplementation of L-arginine to milk replacer improved growth of newly born piglets. Sows are also benefitted by dietary supplementation of L-arginine. Improved blood flow to placenta and mammary gland increased nutrient supply to fetuses for growth and to mammary gland for milk production. Sows with supplemental L-arginine had reduced number of dead born piglets and increased milk production during early stage of lactation.

## 3.8 ENZYMES

Rapid increase in the cost of feed ingredients caused such a big obstacle to swine producers. Shortage of good quality feed grains pushed swine producers to include grain sources that are not in a best quality with indigestible nutrients and with potential anti-nutritional factors and mycotoxins. This caused decrease in production efficiency due to reduced utilization of nutrients in the feed by pigs. Use of exogenous microbial enzymes has greatly been increased in feeding swine. Most of exogenous enzymes are produced by bio-fermentation to help digestion of specific nutrients or compounds in the feed or feed grains. Thus specific enzymes can be supplemented to a diet with specific ingredients with potential problems. One of popular examples is the use of phytase to hydrolyze phytic acid structure releasing phosphorus as well as other minerals.

There are another types of exogenous enzymes that became popular. Carbohydrases or more often called NSPases (non-starch polysaccharidases) helps efficiency of energy utilization of feed grains by pigs. Application of exogenous NSPases should consider types of feed grains used in the diet because types of NSPs are different depending on feed ingredients. For corn-soybean meal based feed that are most widely used in Americas and Asian countries, types of NSPs that interfere energy digestibility are galactosides, mannosides, soybean oligosaccharides.

Growing-finishing pigs can also be benefitted by the use of NSPases as their ability to utilize energy in the feed improves. Swine producers can reduce energy content in the diets without sacrificing growth performance when exogenous NSPases are supplemented. Use of NSPases in lactation diets helped sows often possessing a limited voluntary nutrient intake to preserve body tissues without reducing milk production as their energy digestibility improved.

As previously described, exogenous microbial enzymes produced by bio-fermentation has made a great contribution to swine production by improving production efficiency.

## 3.9 YEAST METABOLITES

There have been many types of probiotics and prebiotics introduced as potential alternatives to the use of supplemental antibiotics as growth promoters. These probiotics and prebiotics mainly target reducing pH and/or reducing population of pathogenic microorganisms in hind gut. However, lack of consistent research evidences showing benefits of probiotics and prebiotics limits application in swine diets. Yeast metabolites have newly been introduced in swine diets. Yeast metabolites are also called yeast culture that includes concepts of both probiotics and prebiotics. Yeast metabolites are first introduced to dairy producers and known to improve dry matter intake of dairy cows increasing milk production.

Recent research with sows conducted at a large commercial swine operation introduced that yeast metabolites may increase milk production for litter growth.

Yeast metabolites are produced by bio-fermentation of cereal grains by Saccharomyces ceravisiae under anaerobic liquid state condition. Yeast cells would preserve essential metabolites for their survival instead of generating ATPs under an anaerobic condition. Solid state fermentation follows liquid state fermentation encouraging further production of metabolites that can be dried and used in feed application. Yeast metabolites, therefore, include metabolites produced during anaerobic fermentation as well as cell content and cell wall components. Repeatable results are obtained showing dietary supplementation of yeast metabolites improves milk production in sows, even though it is premature to conclude which component in yeast metabolites helps milk production at this moment.

## 3.10 FERMENTED INGREDIENTS

Vegetable protein sources including soybean meals have economical advantages over expensive animal protein sources. However, these vegetable protein sources often contain anti-nutritional factors limiting their application in feeding newly weaned pigs with immature gut structure that can be damaged by anti-nutritional factors. There are several anti-nutritional factors found in vegetable protein sources including trypsin inhibitor, lectin, flatulence producing compounds, phytate, glycinin, β-conglycinin, etc.

Some of anti-nutritional factors can be destroyed by feed processing such as heating and cooking. Newly weaned pigs with limited HCl in stomach and enzymatic secretion in small intestine can have difficulties to digest proteins with complex structures and large molecular weights.

There has been an effort to improve vegetable protein sources by reducing anti-nutritional factors using bio-fermentation technology. Aspergillus oryzae and Bacillus subtilis have traditionally been used preparing soybeans via fermentation. Careful selection of these microorganisms for high activity in specific feed ingredient allowed preparation of novel fermented vegetable protein sources. This fermentation process by Aspergillus oryzae and Bacillus subtilis reduced trypsin inhibitor below a detection limit, reduced phytate content by 4-folds, reduced size of peptides for suitable for digestion, and reduced contents of glycinin and β-conglycinin that are known as allergenic compounds in gut.

Use of these fermented vegetable protein sources in nursery diets improved growth efficiency of newly weaned pigs compared to pigs fed same vegetable protein sources without fermentation. Fermented vegetable protein sources could also replace the use of animal protein sources such as whey protein concentrates and spray dried plasma protein in part without compromising growth performance. It is a great contribution to swine industry applying bio-fermentation technology to prepare relatively inexpensive vegetable protein sources replacing the use of expensive animal protein sources in swine diets.

The Biofermentatio Technology allows for 100X of effective product to be dosed in to a specific environment thereby providing astonishing results in turning around biological wastewater treatment systems in days not weeks, starting up plants in hours not weeks.

ABS is an Enviromental Technology Company with over 20 years experience and 50+ patents. ABS has a proven commercialized technology called Biofermentation™ that can grow thousands of pounds of oil eating microbes at a "Staging Facility" in your county or city, which, when applied, result in "Accelerated Natural Attenuation." This technology has been given a favorable review by BP Horizon Spill Response. The technology has been presented on CNN by an ABS distributor RTG (Todd Brown). Biofermentation™ offers the only solution to scale-up the oil eating microbes necessary to bioremediate marshes and the oil stains left on the beaches after removal of tar balls.

### 3.10.2 What are Microbes?

Microbes are living organisms such as bacteria or fungi. ABS ONLY uses NATURAL STRAINS OF MICROBES, genetically engineered microbes (GEMS) are highly regulated and are not legally allowed to be dispersed in to the environment without specific controls.

### 3.10.3 What do Microbes do?

Microbes eat organic compounds to grow and survive converting these organics to harmless end products of carbon dioxide, water and more microbes. Many microbes are useful for biodegrading organic contamination, while others use hydrogen sulfide or ammonia. The latter microbes can be harnessed to control odors or to enhance nitrification. The process of biodegradation involves a complex series of reactions involving many enzymes, so microbes are essentially enzyme factories, each enzyme having a specific purpose, but with the microbe managing the ultimate outcome.

# 4

# FERMENTERS AND BIOREACTORS

## 4.1 INTRODUCTION

Bioreactor and fermentor are two types of vessels that allow the occurrence of biochemical reactions by providing the optimum conditions for reactions. A bioreactor is a device which facilitates various types of biochemical reactions while fermentor is a device that facilitates fermentation. Fermentor is a type of bioreactor. Bioreactors can be used in the production of pharmaceutical products, vaccines or antibodies while fermenters are used in the production of ethanol and lactic acid. The main difference between bioreactor and fermentor is the type of biochemical reaction they carry out.

## 4.2 FERMENTERS

Fermentor is a type of bioreactor, which uses fungi or bacteria for the fermentation of ethanol or lactic acid. Hence, a fermentor operates under anaerobic conditions. Also, fermentor is capable of providing the optimal growth conditions such as temperature to microorganisms.

**Fig. 1. Fermenters**

Based on the type of cultures used in a fermentor, it can be classified into three as batch, fed-batch or continuous. Also, two type of fermentations can occur in a fermentor: surface fermentation by microorganisms in a solid medium and submerged fermentation by microorganisms in a liquid medium.

## 4.3 BIOREACTORS

A bioreactor is a closed system used for the processing of a biochemical reaction. It supports either the growth of cells such as mammalian or insect in a culture or production of a secondary metabolite such as pharmaceutical products, antibodies or vaccines. Fermentor is a type of bioreactor that uses fungal or bacterial cells for fermentation. Thus, depending on the purpose, bioreactors can be classified into two types; suspended growth bioreactors, which produce secondary metabolites, and biofilm bioreactors, which process cell cultures.

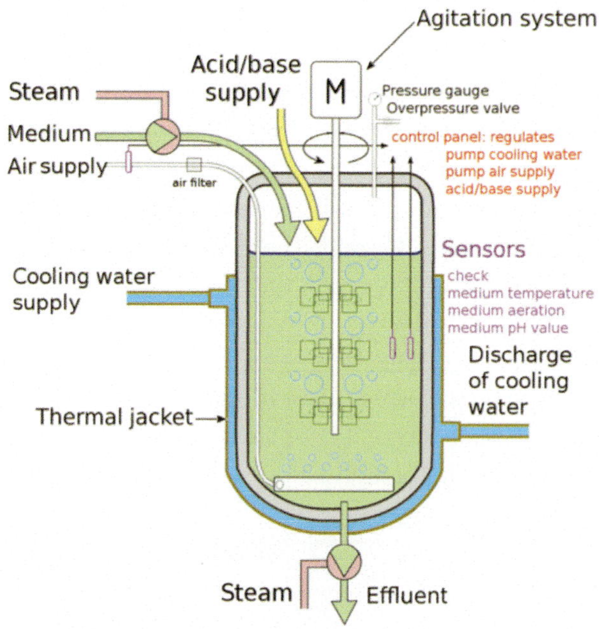


**Fig. 2. Bioreactors**

The six bioreactor designs are continuous stirred tank bioreactors, bubble column bioreactors, airlift bioreactors, fluidized bed bioreactors, packed bed bioreactors, and photo-bioreactors.

A bioreactor is a device in which the microorganisms are cultivated and motivated to form the desired products by maintaining optimum conditions for growth and metabolic activity. A fermenter refers to the device used for the cultivation of prokaryotic cells e.g. bacteria, fungi etc. whereas a bioreactor is used for growing eukaryotic cells. A typical conventional bioreactor has cylindrical vessel with domed top and bottom generally made up of stainless steel. The reaction vessel which is surrounded by a jacket, is provided with a sparger at the bottom through which air (or other gases such as $CO_2$ and $NH_3$ (for pH maintenance) can be introduced. The reaction vessel also has side ports for pH, temperature and dissolved $O_2$ sensors. The agitator shaft is connected to a motor at the bottom. Above the liquid level of the reaction vessel, connections for acid, alkali, antifoam chemicals and inoculum are located. The bioreactor is designed to work at very high temperatures (150-1800C), high pressure (377-412 kPa) and also to withstand vacuum which prevents its collapse while cooling.

### 4.3.1 Bioreactor Design

- The design and mode of operation of a bioreactor are based on the production of an organism, optimum conditions required for desired product formation, product value, and its scale of production.
- A good bioreactor design will help to improve productivity and provide higher quality products at lower prices.
- A bioreactor is a device that consists of various features such as an agitator system, an oxygen delivery system, a foam control system, and a variety of other systems such as temperature & pH control system, sampling ports, cleaning, and sterilization system, and lines for charging & emptying the reactor.
- The material used for the construction of a bioreactor must have the following important properties:

  It should not be corrosive.

  It should not add any toxic substances to the fermentation media.

  It should tolerate the steam sterilization process.

  It should be able to tolerate high pressure and resist pH changes.

  The sizes of the bioreactor vary widely depending on the application.

Some bioreactors are designed for small scale fermenters and some for large scale industrial applications from the microbial cell (few $mm^3$) to shake flask (100-1000 ml) to the laboratory-scale fermenter (1 – 50 L) to pilot level (0.3 – 10 $m^3$) to plant scale (2 – 500 $m^3$) for large volume.

### 4.3.2 Bioreactor Principle

- The bioreactor is the heart of any biochemical process as it provides an environment for microorganisms to obtain optimal growth and produce metabolites for the biotransformation and bioconversion of substrates into desirable products.
- The reactors can be engineered or manufactured based on the growth requirements of the organisms used.

- Reactors are machines that can be made to transform biological-based materials into desirable products.
- They can be used for the production of various enzymes and other bio-catalysis processes.

## 4.4 PARTS OF THE BIOREACTOR AND THEIR FUNCTION

- These reactors have been designed to maintain certain parameters like flow rates, aeration, temperature, pH, foam control, and agitation rate.
- The number of parameters that can be monitored and controlled is limited by the number of sensors and control elements incorporated into a given bioreactor.
- Other factors should be kept in mind before designing a fermenter as described below:

**1. Fermenter Vessel**

- A fermenter is a large cylinder closed at the top and bottom connected with various pipes and valves.
- The vessel is designed in such a way that it allows to work under controlled conditions.
- Glass and stainless steels are two types of fermenter vessels used.
- The glass vessel is usually used in small-scale industries. It is non-toxic and corrosion-proof.
- Stainless steel vessel is used in large scale industries. It can resist pressure and corrosion.

**2. Heating and Cooling Apparatus**

- The fermentor vessel's exterior is fitted with a cooling jacket that seals the vessel and provides cooling water.
- Thermostatically controlled baths or internal coils are generally used to provide heat while silicone jackets are used to remove excess heat.
- A cooling jacket is necessary for sterilization of the nutrient medium and removal of the heat generated during fermentation in the fermentor.

**3. Aeration System**

- An aeration system is one of the very important parts of a fermentor.
- It is important to choose a good aeration system to ensure proper aeration and oxygen availability throughout the culture.
- It contains two separate aeration devices to ensure proper aeration in a fermentor.
- The stirring accomplishes two things:

- It helps to mix the gas bubbles through the liquid culture medium and
- It helps to mix the microbial cells through the liquid culture medium which ensures the uniform access of microbial cells to the nutrients.

**4. Sealing Assembly**

- The sealing assembly is used for the sealing of the stirrer shaft to offer proper agitation.
- There are three types of sealing assembly in the fermenter:
- Packed gland seal
- Mechanical seal
- Magnetic drives

**5. Baffles**

- The baffles are incorporated into fermenters to prevent a vortex improve aeration in the fermenters.
- It consists of metal strips attached radially to the wall.

**6. Impeller**

- Impellers are used to provide uniform suspension of microbial cells in different nutrient mediums.
- They are made up of impeller blades attached to a motor on the lid.
- Impeller blades play an important role in reducing the size of air bubbles and distribute them uniformly into the fermentation media.
- Variable impellers are used in the fermenters and are classified as follows.
- Disc turbines
- Variable pitch open turbine

**7. Sparger**

- A sparger is a system used for introducing sterile air to a fermentation vessel. It helps in providing proper aeration to the vessel.
- The sparger pipes contain small holes of about 5-10 mm, through which pressurized air is released.
- Three types of sparger are used
- Porous sparger
- Nozzle sparger
- Combined sparger–agitator

**8. Feed Ports**

- They are used to add nutrients and acid/alkali to the fermentor.
- Feed ports are tubes made up of silicone.
- In-situ sterilization is performed before the removal or addition of the products.

**9. Foam-Control**

- The level of foam in the vessel must be minimized to avoid contamination, this is an important aspect of the fermentor.
- Foam is controlled by two units, foam sensing, and a control unit.
- A foam-controlling device is mounted on top of the fermentor, with an inlet into the fermentor.

**10. Valves**

- Valves are used in the fermentor to control the movement of liquid in the vessel.
- There are around five types of valves are used, that is,
- Globe valve,
- Butterfly valve,
- A ball valve, and
- Diaphragm valve.
- A safety valve is built-in in the air and pipe layout to operate under pressure

**11. Controlling Devices for Environmental Factors**

- A variety of devices are utilized to control environmental elements like temperature, oxygen concentration, pH, cell mass, essential nutrient levels, and product concentration.

**12. Use of Computer in Fermenter**

- For an efficient process, monitoring, and data collecting, fermentors are generally coupled with modern automated and semi-automated computers and databases.

## 4.4.1 Types of Bioreactors

Depending on the design of the reactor, the bioreactors are of following types:

**(a) Continuous stirred tank bioreactors:** These bioreactors have a cylindrical vessel with motor driven central shaft which gives support to one or more agitators (impellers). The shaft is fitted at the bottom of the bioreactor. The diameter of the impeller is usually one third of the vessel diameter. The impellers are available in different designs like-Rustom disc, concave bladded, marine propeller etc. In stirred tank reactors, the air is added to the culture medium under pressure through a device called sparger. The sparger along with the impellers (agitators) enables better and efficient gas distribution through out in the vessel. The advantages of using stirred tank reactors are: the efficient transfer of gas to growing cells which keeps the growth of cells in healthy limits, stirring ensures good mixing of the contents, the operating conditions are flexible and the bioreactors are easily available which makes them commercially viable products.

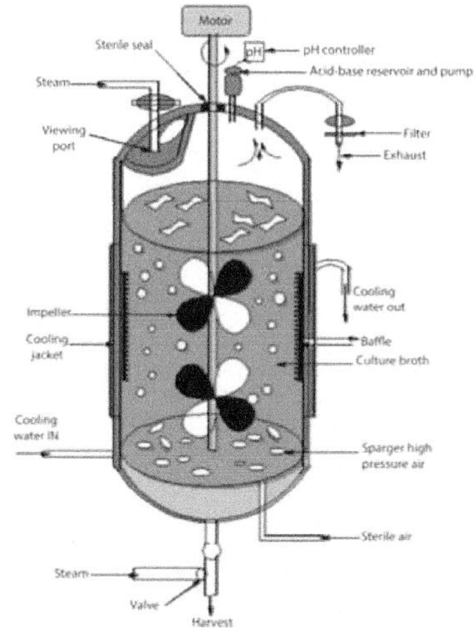


**Fig. 3. Continuous stirred tank fermentor**

**(b) Bubble column bioreactors:** In these bioreactors, the gas or air is introduced at the base of the column through perforated pipes or plates, or metal microporous spargers. The vessel used for bubble column bioreactors is usually cylindrical with an aspect ratio (height o diameter ratio) of 4-6. The rate of flow of gas affects the $O_2$ transfer and mixing.

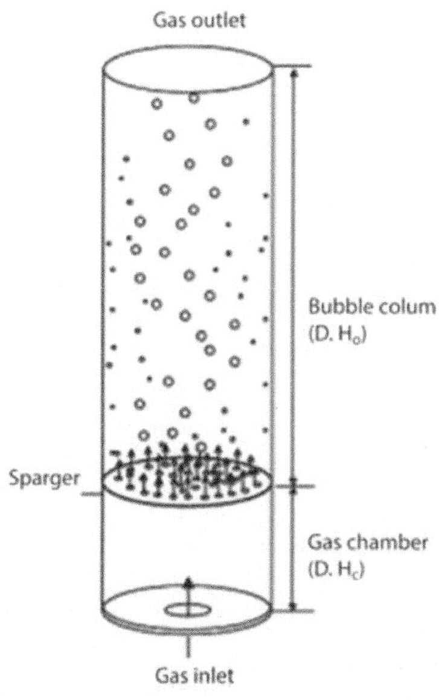


**Fig. 4. Bubble column fermentor**

**(c) Airlift bioreactors:** Airlift bioreactors are commonly used for aerobic bioprocessing technology. In the airlift bioreactors, the medium of the vessel is divided into two interconnected zones by means of baffle or draft tube. The air/gas is pumped into one of the two zones referred to as 'riser' and the other zone that receives no gas is known as 'downcomer'. The dispersion flows up the riser zone while the down flow occurs in the downcomer.

These bioreactors have a single container with a central draft tube that creates interior liquid circulation channels which keeps the volume and circulation at a fixed rate for fermentation.External loop airlift bioreactor-These have an external loop to keep the liquid in circulation through separate independent channels. The modifications can be made in these bioreactors depending on the requirements of different fermentation processes.

**Two stage airlift bioreactors:** These bioreactors have two bioreactors which are basically used for the temperature dependent formation of products. The growing cells from one bioreactor (maintained at temperature 300°C are pumped into another bioreactor (at temperature 420°C). This is done because it is very difficult to increase the temperature quickly from 300°C to 420°C in the same vessel. The cells are grown in the first bioreactor and with the help of the fitted valves and a transfer tube and pump, they are transferred into the second bioreactor, where the actual bioprocessing takes place. Tower bioreactors - In this type of bioreactor, a high hydrostatic pressure is generated at the bottom of the reactor which increases the solubility of $O_2$ in the medium. Since the top is expanded, the pressure is reduced which helps in the expulsion of $CO_2$. The cycle completes with the medium flowing back into the downcomer. The advantage with Tower bioreactor is that it has high aeration capacities without having moving parts.

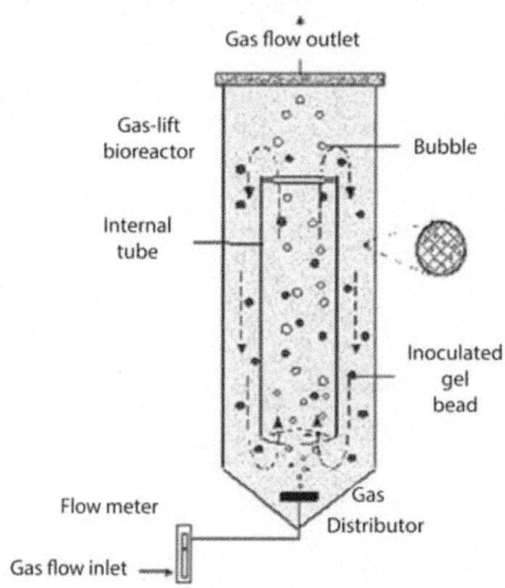


**Fig. 5. Fermentor**

**(d) Fluidized bed bioreactors:** These bioreactors are mainly suitable to carry out reactions involving fluid suspended biocatalysts such as immobilized enzymes, immobilized cells, microbial flocs etc. The design of the bioreactors is such that the top is extended and the reaction column is narrow which retains the solids in the reactor and allows the liquids to flow out. To maintain an efficient operation of fluidized beds, gas is sparged to create a suitable gas-liquid-solid fluid bed. The recycling of the liquid ensures continuous contact between the reaction contents and biocatalysts which increases the efficiency of bioprocessing.

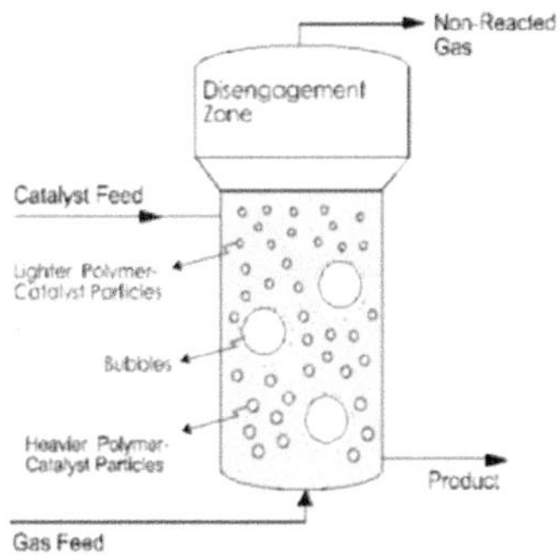


**Fig. 6. Figure: Fluidized-bed fermentor.**

**(e) Packed bed bioreactors:** A packed bed bioreactor consists of a bed of solid particles, with biocatalysts on or within the matrix of solids, packed in a column. The solids are generally porous or non-porous gels which maybe compressible or rigid in nature. The nutrient broth continuously flows over the immobilized biocatalyst and the products are released into the fluid from where they are removed. However, due to poor mixing, it is difficult to control the pH of packed bioreactors by the addition of acid or alkali.

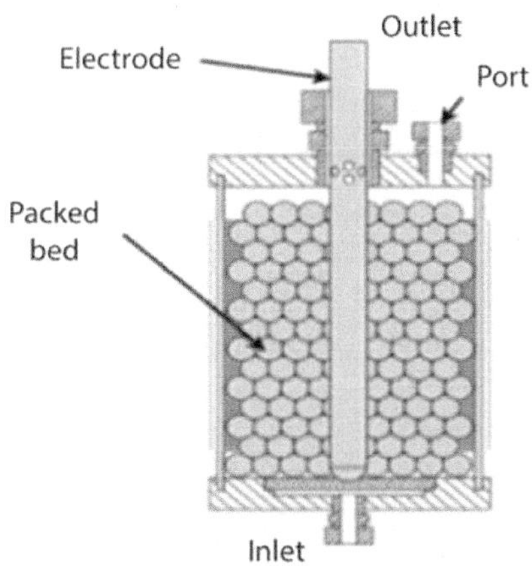


**Fig. 7. Packed Bed Bioreactor**

**(f) Photobioreactors:** These bioreactors are specialized for fermentation that can be carried out either by exposing to sunlight or artificial illumination. The photobioreactors are made up of glass or transparent plastic which are the solar receivers. The cell cultures are circulated through the solar receivers by using centrifugal pumps or airlift pumps. These bioreactors work in the temperature which ranges from 25-400°C. In these bioreactors, the microorganisms e.g. microalgae, cyanobacteria etc. grow during the day time while the products (e.g. beta-carotene, asthaxanthin) are produced during the night.

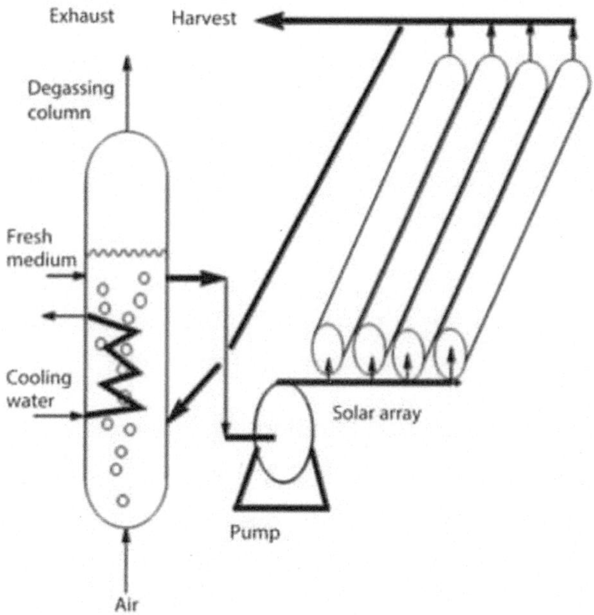


A tubular photobioreactor with parallel run horizontal tubes.

**Fig. 8. Photobioreactor**

### 4.4.2 Solid Substrate Fermentation (SSF)

In some biotechnological processes, the growth of the microorganisms is carried out on solid substrates more or less in the absence of free water. Only approximately 15% of moisture is present which is essential for solid-substrate fermentation. Cereal grains, wheat bran, sawdust, wood shavings etc are some of the commonly used solid substrates for SSF. The technique of SSF is used for the production of edible mushrooms, cheese, soy sauce, and many enzymes and organic acids. It is carried out as non-aspetic process and therefore saves sterilization costs. The bioreactors used in this type of fermentation process have simple designs with simple aeration process and effluent treatment. The yield of the products is very high, at low energy expenditure. However, in this process only microorganisms, that can tolerate only low moisture content can be used. It also difficult to monitor $O_2$ and $CO_2$ levels in SSF. The slow growth of microorganisms, also become a limiting factor for product formation.

## 4.5 WORKING OF A BIOREACTOR

In the operation of a bioreactor, there are a few steps that are followed:

**(a) Sterilization:** The most important requirement is to maintain aseptic or sterile conditions for aseptic fermentation. In order to achieve this, the air supplied during fermentation, he growth medium and the bioreactor it self and all it's accessories are sterilized. There are two methods of sterilization that are followed:

**(1) In situ sterilization:** In this, the bioreactor is filled with the required medium followed by injection of pressurized steam into the jacket or coil surrounding the reaction vessel. The whole system is heated to about 1200°C and maintained at this temperature for about 20 minutes. However, this method is not energy efficient and prolonged heating destroys the vitamins and precipitates the components of the medium.

**(2) Continuous heat sterilization:** In this, the empty bioreactor is first sterilized by injecting pressurized steam and the medium is rapidly heated to 1400°C for a short period again by injecting the pressurized steam. This is an energy efficient method and also does not precipitate the medium components.

**(b) Aeration:** Oxygen is stored in compressed tanks and is introduced at the bottom of the bioreactor through a 'sparger'. Aeration of the fermentation medium supplies oxygen to the production microorganisms and remove carbon dioxide from the bioreactor. The gases released during the fermentation accumulate in the headspace and then pass out through an air outlet. The headspace is a vacant space on the upper part of the bioreactor and is generally about 20% of it's volume. The air-lift system of aeration involves sparging of air done at the bottom of the fermenter with an upward flow of air bubbles. The aeration capacity of the system depends on the air-flow rate and the internal pressure. The stirred system of aeration involves increasing the aeration capacity by stirring using impellers driven by motor. The aeration capacity of the fermenter depends on the rate of stirring, rate of air flow and internal pressure.

**(c) Inoculation and sampling:** The sterilized bioreactors with growth medium are inoculated with the production organisms. The size of the inoculum is generally 1-10% of the total volume of the medium. During the fermentation process, the samples are regularly withdrawn to check contamination and to measure the amount and quantity of product formed.

**(d) Control systems:** Various factors like- pH, temperature, dissolved oxygen, adequate mixing, concentration of the nutrients, foam formation etc are continuously monitored to maintain optimal growth environment in the bioreactor. Very sensitive sensors are available which carry out automated monitoring of these variables. The ideal pH range for optimal growth of microorganisms is between 5.5-8.5. The pH changes due to the release of metabolites into the medium by the growing microorganisms. The required pH level is maintained by adding acid or alkali followed by thorough mixing of the medium components. The optimal temperature is maintained by using the heating and cooling systems fitted in the bioreactor. Continuous monitoring of dissolved oxygen concentration is also a must for the optimal bioreactions. The oxygen is sparingly soluble in water (0.0084 g/l at 250°C) and is introduced into the bioreactor as the air bubbles.

The concentration of the nutrients is also important because the limiting concentrations of nutrients helps in the optimal product formation and the high nutrient concentrations have inhibitory effect on the microbial growth. Another important factor to control is the "foam formation". The protein rich media is used in industrial fermentation which leads to 'froth' or 'foam formation' on agitation during aeration. Some antifoam chemicals lower the surface tension of the medium and causes the foam bubbles to collapse. Mineral oils with silicone or vegetable oils are also used as antifoam agents. The bioreactors can also be fitted with mechanical foam control devices which break the foam bubbles and throw back into the fermentation medium. The proper and continuous mixing of the microbial culture is very important to maintain optimal levels of oxygen in the nutrient medium and to prevent the accumulation of toxic metabolic products.

**(e) Cleaning:** After the completion of the fermentation process, the products are 'harvested' and the bioreactor is prepared for the next round of fermentation after cleaning technically referred to as 'turn around'. The cleaning of the bioreactors is carried out by using high-pressure water jets from the nozzles fitted into the reaction vessel. In order to maintain the cost effectiveness of the bioreactor, the time taken for turn around which is known as 'down time', is kept as short as possible.

## 4.6 DOWNSTREAM PROCESSING (DSP)

The extraction and purification of a biotechnological product from fermentation is referred to as downstream processing. The methods adopted for downstream processing depends on the nature of the end products, it's concentration, stability, and the degree of purification required. Both intracellular metabolites as well as extracellular metabolites are isolated by DSP. The intracellular metabolites are products located within the cells e.g. vitamins, enzymes etc. The extracellular metabolites are the products present outside the cells e.g. most antibiotics, amino acids, alcohol, citric acid, enzymes like amylases, proteases etc. some products like vitamin B12, flavomycin etc are present both as intracellular as well as extracellular products.

Downstream processing involves a number of steps which are as follows:

**(A) Solid-liquid separation:** The first step is to separate whole cells and other insoluble substances from the culture broth. This is done by using several methods:

1. **Flotation:** The process of aeration involves the bubbling of gas in to the liquid broth. The cells and other solid particles get adsorbed on gas bubbles which form a foamy layer which is collected and removed.

2. **Flocculation:** The cells or cellular debris form large aggregates and settle down which can be easily removed. Some flocculating agents like inorganic salt, organic poly-electrolyte, mineral hydrocolloid etc are often used to achieve appropriate flocculation.

3. **Filtration:** This is the most commonly used technique to separate the biomass and culture filtrate. The rate of filtration depends on many factors such as the size of the organism, presence of other organisms, viscosity of the medium, and temperature. Several filters like depth filters, absolute filters, rotary drum vacuum filters, membrane filters etc. are used. There are three major types of filtration processes used depending on the size of the particles- microfiltration, ultrafiltration, and reverse osmosis.

4. **Centrifugation:** The centrifugation is mostly used for separating solid particles from liquid phase. The technique of centrifugation is based on the principle of density differences between the particles to be separated and the medium. In the bioreactors, the continuous flow industrial centrifuges are used where there is a continuous feeding of the slurry and collection of clarified fluid. The solid deposits are intermittently removed. Various models of centrifuges used are: Tubular bowl centrifuge, Disc centrifuge, Multichamber centrifuge, Scroll centrifuge or decanter etc.

**(B) Release of intracellular products:** The biotechnological products that are located with in the cells like vitamins, enzymes, etc are released in an active form for their further processing and isolation. The microorganisms or cells can de disintegrated or disrupted by physical, chemical, or enzymatic methods depending on the nature of the cells.

(1) **Physical methods of cell disruption:** These methdos are (i) Ultrasonication, (ii) Osmotic shock (used for releasing hydrolytic enzymes and binding proteins from gram-negative bacteria), (iii) Heat Shock treatment, (iv) High pressure homogenization, (v) Impingement which involves hitting a stationary surface or a second stream of suspended particles with a stream of suspended cells at high velocity and pressure. Microfluidizer is a device developed on the basis of the principle of impingement. (vi) Grinding with glass beads where the cells mixed with glass beads are subjected to a very high speed in a reaction vessel. The cells break as they are forced against the wall of the vessel by the beads

(2) **Chemical methods:** Treatment with alkalies, organic solvents, and detergents lyse the cells to release the content. Alkali treatment is used for the extraction of some bacterial proteins. The organic solvents like methanol, ethanol, isopropanol, butanol etc also disrupt the cells. The organic solvent, toluene, which is commonly used, dissolves membrane phospholipids and creates the membrane pores to release intracellular contents. The ionic detergents denature the membrane proteins and lyse the cells e.g. cationic-cetyl trimethyl ammonium bromide or

anionic-sodium lauryl sulfate. Non-ionic detergents are less reactive and also affect the purification steps.

(3) **Enzymatic methods:** Lysozyme is the most commonly used enzyme which hydrolyses beta-1,4-glycosidic bonds of the mucopepide in bacterial cell walls e.g. gram positive bacteria. This enzyme is commercially available produced from hen egg white. For gram-negative bacteria, lysozymes are in use with EDTA to break the cells. When the cell wall gets digested by lysozyme, the periplasmic membrane breaks due to osmotic pressure, which releases the intracellular contents. Glucanase and mannanase along with proteases lyse the cell wall of yeast.

**(C) Concentration:** The biological products are concentrated by getting rid of water which is present in the filtrate. Depending on the nature of the desired products and the cost effectiveness, the techniques used to concentrate the biotechnological products are evaporation, liquid-liquid extraction, membrane filtration, precipitation, and adsorption.

### 4.6.1 Similarities Between Bioreactor and Fermentor

- Bioreactor and fermentor are two types of vessels that facilitate a specific type of biochemical reaction.
- The main function of them is to provide agitation, aeration, sterility, regulation of factors like temperature, pH, pressure, nutrient feeding, liquid level, etc. and allow withdrawal of cells or the medium.
- These vessels are generally cylindrical and made up of stainless steel.
- hey operate on a large scale.
- Both are closed systems.

Bioreactor refers to an apparatus in which a biological reaction or process is carried out, especially on an industrial scale whereas fermentor refers to the container in which fermentation takes place.

# 5

# FOOD FERMENTATION AND MICRO-ORGANISMS

## 5.1 INTRODUCTION

Within the fermentation industry, microorganisms are used for the production of specific metabolites such as acids, alcohols, enzymes, antibiotics, carbohydrates. Major fermentation microorganisms include lactic acid bacteria (LAB), moulds and yeasts. In particular, LAB are the major microflora involved in fermented dairy products, vegetable, and sourdough fermentation, and (mainly lactobacilli and pediococci) are part of the starter cultures used in meat fermentation to produce desirable acids and flavour compounds. Industrial control of fermentation processes requires up-to-date knowledge of the physiology, metabolism and genetic properties of such microorganisms. Equally important is the knowledge of their impact on food quality, safety and shelf-life. The character of food fermentation, however, is more complex. Agricultural products of animal or vegetable origin are fermented by either the indigenous microflora or an added starter culture to improve or obtain shelf life, nutritional value, health benefit, flavour or texture. Unlike the Western world, Asia has developed many foods based on vegetable proteins using fungi, often in a solid-state fermentation process. In Europe and the US the main focus of food fermentation has been on food preservation by means of acid fermentation, whereas properties such as taste, nutritional value and health effect are more important in Asian (fungal) fermentation. Several authors have written comprehensive reviews about fermented foods.

"What microbiology has illuminated is that all the plants and animal products that make up our food are inevitably populated by a diversity of microorganisms," emails Sandor Katz, self-described "fermentation revivalist," teacher and author of several books about fermentation, including "Wild Fermentation," "The Art of Fermentation," "Fermentation as Metaphor," and "Fermentation Journeys."

"Fermentation involves encouraging the growth of some microorganisms — ones that make our food more stable, more delicious, more nutritious, safer, less toxic or alcoholic — and thereby prevent the growth of the ones that can decompose our food or make us sick," says Katz. "Certain ferments, such as strong cheeses, are edgy in that some people find them compellingly delicious, while others think they are disgusting and are reminded of decomposition and death. Science offers us no sharp dividing line between fermentation and spoilage. In most cases it is obvious, but in certain cases it is culturally determined and subjective."

Pretty much every culinary tradition in the world uses fermentation in some way or another. Humans began using fermentation to brew beer, preserve food, make their food tasty and render toxic things edible millennia before Louis Pasteur proved that living cells were responsible for fermentation.

"The oldest examples of fermentation in the archeological record are from about 10,000 years ago in China, though I would argue that that tells us more about the history of pottery than the history of fermentation," says Katz. "Presumably the cultural practice of fermentation is older than that, using pits in the ground, gourds, animal membranes, wood or other biodegradable materials."

According to Katz, it's no surprise that fermentation can be traced back to around the same time as humans began developing communities around domesticating plants and animals. In order for it to make sense to invest all our time and energy into growing food, humans would have had to come up with a strategy for storing and preserving food so that the harvest of a few weeks could last the community many months.

Also, basically every culture in the world has figured out how to ferment sugars into alcohol. It's possible our primate ancestors understood the party-inducing effects of eating a bunch of fermented fruit they found on the ground. All humans had to do was figure out how to do it on purpose.

The word "ferment" comes from the Latin word for boiling or rising, which you'll understand if you've ever seen beer brewing or bread dough rising — both signs of fermentation in action. A biologist would probably say fermentation is the anaerobic production of energy, meaning that microorganisms that don't need oxygen to accomplish metabolization to fuel their bodies transform nutrients into energy in the absence of oxygen. This is true for some fermented foods — for instance, sauerkraut and alcohol are both products of the hard work of anaerobic bacteria. However, other foods and beverages that we consider "ferments" are made by aerobic microorganisms that use oxygen to do their business: kombucha, vinegars and miso are a few.

The applied science of zymology, or zymurgy, is the study of how microorganisms, whether they use oxygen to metabolize or not, ferment the stuff around us — what the biochemical processes are, which specific organisms are doing it, and how this works all over the world, including in our foods.

There are three basic types of fermentation that can happen in the foods and drinks we enjoy every day:

- Alcoholic fermentation has been a fermentation fan favorite for millennia — it uses yeasts to convert sugars to ethanol, carbon dioxide and other byproducts.
- Acetic acid fermentation happens after alcoholic fermentation ends, and the result is usually vinegar.
- Lactic acid fermentation is the magic behind so many of our favorite foods, from sourdough bread to kimchi to cheese and yogurt. Lactic acid bacteria digest simple carbohydrates to produce lactic acid — a process that turns sweet things sour, which is what gives pickles their characteristic kick.
- Not all fermented foods and beverages do the same things in our bodies, but according to Katz, they can help us out in a few important ways:

**Predigestion:** The microorganisms that ferment the food essentially break down things like carbohydrates, proteins and chemical bonds before we eat it, in some cases making it more easily digested and bioavailable. "Even nutrients which create problems for many people, such as lactose or gluten, can be broken down by fermentation," says Katz.

**Detoxification:** Some foods can only be made edible through fermentation. The microorganisms take potentially toxic (to us) compounds and digest them into harmless or even beneficial forms, rendering foods that could otherwise be poisonous or irritating safe to eat. One example of this can be found in cassava tubers grown in some regions of the world — they contain high levels of cyanide when they're harvested and require fermentation to make them safe to eat.

**Nutrient enhancement**: Fermentation generates additional B vitamins in most foods, K vitamins in some. Some metabolic byproducts we can think of as micronutrients are just beginning to be investigated.

**Probiotics:** "I think of the bacteria themselves as the most profound benefit of fermented foods, in that they exist in great diversity and can help to restore biodiversity in the gut that has diminished thanks to chemical exposure as well as modern diets high in overprocessing and low in fiber," says Katz. The trillions of bacteria that live in and on our bodies are physiologically important, playing important roles in digestion, immune function, synthesizing nutrients and regulating chemicals that impact many different organs, including our brains. Eating live fermented foods, especially a variety of them, can help to restore biodiversity in the gut and potentially improve digestion, immune function and more.

- To understand the scope of fermented foods in nutrition and health, it is necessary to acknowledge the wide diversity of microorganisms used for fermented food production. The most common fermented foods and beverages require lactic acid bacteria (LAB), AAB, bacilli or other bacteria, yeasts, or filamentous fungi. These microorganisms were among the very first to be isolated and characterized by Pasteur, Lister and other early microbiologists and have long served as model organisms in biology and as a source of industrial chemicals and bioactive molecules. More recently, they were integral to the discovery and application of CRISPR technology.
- LAB are a group of Gram-positive, non-spore forming, aerotolerant bacteria that are phylogenetically positioned within the Firmicutes phylum, predominantly in the order Lactobacillales. They are among the most important and widely used microorganisms in food fermentations, serving essential functions in fermented dairy, meat, cereal and vegetable products. LAB include the reclassified members of the Lactobacillaceae or Lactobacillus genus complex41 and numerous other taxa, including species of Lactococcus and Tetragenococcus associated with milk and soy sauce fermentations, respectively. Besides LAB, particular species of Bacillus and AAB are solely responsible for some fermented foods or have important supporting roles as is the case for Staphylococcus, Enterococcus, Brevibacterium and Propionibacterium in sausage and cheese fermentations. Among the fungi, ethanol-producing yeasts, usually species of Saccharomyces, are used for bread, beer, wine and various alcoholic fermentations. Interestingly, the domestication of Saccharomyces cerevisiae strains and their adaptation to a range of fermentation substrates and environments has led to the formation of distinct lineages associated with particular products.
- Similar domestication events are also likely responsible for the widespread use of atoxigenic filamentous fungi46. Penicillium, Aspergillus and Rhizopus are among the moulds commonly used for fermented dairy, meat and soy products and include proteinase, lipase and amylase-producing strains. As described later, many food fermentations involve microbial communities consisting of multiple genera and taxa.
- Considerable progress has been made towards understanding the function of individual microorganisms in fermented food production and then using that information to improve products and strains. Phylogenomic analyses have

shown that, despite their general biochemical and physiological similarities, wine, beer and bread yeast strains evolved independently based on habitat and geography as well as through human-driven domestication. Since the twentieth century, pure starter cultures have been developed to provide consistency and convenience and to accommodate large-scale industrial fermentations. Typically, only one or two microbial strains (for example, bread, yoghurt, cheese) are necessary to initiate those fermentations51. Although technological performance properties remain one of the main criteria, the isolation and development of new strains increasingly relies on relevant genomic information and on the application of available molecular tools.

- Culture-dependent methods remain the gold standard for the strain-level characterization of fermentation microbiota; however, these methods are increasingly complemented by holistic, meta-omics methods. Molecular approaches have shown that fermented foods are frequently dependent on complex, multi-kingdom, microbial communities functioning in concert via dynamic succession processes. However, despite this complexity, the presence of a so-called core microbiota are often apparent in a wide range of fermented foods. Provided that the starting materials are generally the same, spontaneous fermentations typically result in products that contain very similar microorganisms (even the same species), regardless of provenance. For example, fermentations of cabbage and other green leafy vegetables are all initiated by Leuconostoc mesenteroides followed by Lactiplantibacillus species and Levilactobacillus brevis, independent of whether the product is called Sauerkraut (Europe and North America), Kimchi (Korea), Suan-cai (China) or Sinki (Nepal). This highly reproducible succession of fermentation microbiota in spontaneous vegetable fermentations, whereby the assembly of fermentation microbiota is limited by dispersal, reflects the stable association of these organisms with the raw materials (Fig. 1). Similar reproducible successions occur in fungi-fermented foods. Collectively, these and other observations suggest that selective and competitive pressures drive microbiome assembly and succession dynamics and provide a basis for predicting the outcome of food fermentations. Thus, provided that the raw materials and environmental conditions are consistent with the typical practices used for making that food and that salt concentrations, pH, atmosphere or other expected control measures are in place, unpredictable events, which constitute fermentation failure, are relatively rare (Fig. 1). In the absence of those conditions and control measures, food fermentations could result in inferior or unsafe products.

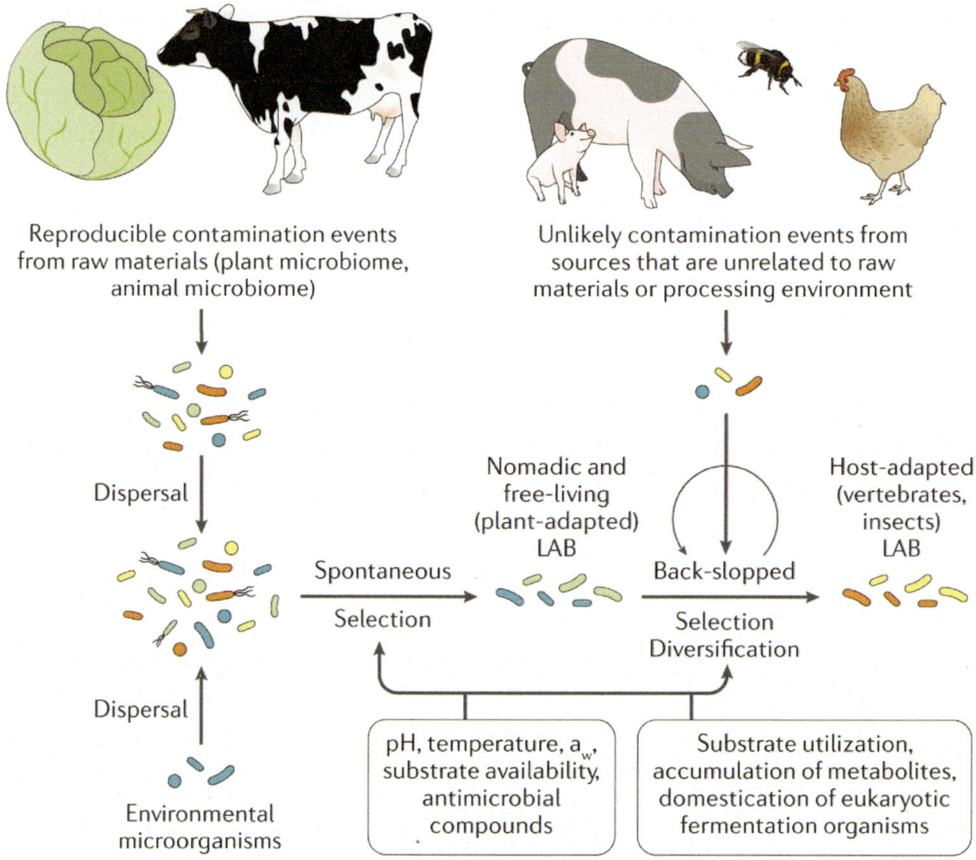


**Fig. 1. Processes that Determine Community Assembly in Traditional Fermented Foods.**

- The conditions established during traditional and industrial fermentations provide a basis for controlling and manipulating autochthonous and allochthonous microorganisms. Microbial communities in spontaneous food fermentations are determined by dispersal and selection. In most spontaneously fermented foods, plant-associated or animal-associated microorganisms are dominant. Back-slopping of fermented foods eliminates dispersal limitation, and selection is the major principle that determines community assembly. Among lactic acid bacteria (LAB), nomadic and free-living species are dominant in spontaneous food fermentations while host-adapted species dominate many back-slopped fermentations. Speciation and domestication have been demonstrated for eukaryotic food fermenting organisms, including Saccharomyces cerevisiae and Aspergillus oryzae, but not for bacteria. If comparable raw materials and fermentation protocols are employed, community assembly in fermented foods is reproducible at the genus level (spontaneous food fermentations) or even at the species level (back-slopped food fermentations).

## 5.2 MICRO-ORGANISMS INVOLVED IN FOOD FERMENTATIONS

The most common groups of micro-organisms involved in food fermentations are:

- Bacteria
- Yeasts
- Moulds

### 5.2.1 Bacteria

Several bacterial families are present in foods, the majority of which are concerned with food spoilage. As a result, the important role of bacteria in the fermentation of foods is often overlooked. The most important bacteria in desirable food fermentations are the lactobacillaceae which have the ability to produce lactic acid from carbohydrates. Other important bacteria, especially in the fermentation of fruits and vegetables, are the acetic acid producing acetobacter species.

### 5.2.2 Yeasts

Yeasts and yeast-like fungi are widely distributed in nature. They are present in orchards and vineyards, in the air, the soil and in the intestinal tract of animals. Like bacteria and moulds, yeasts can have beneficial and non-beneficial effects in foods. The most beneficial yeasts in terms of desirable food fermentation are from the Saccharomyces family, especially S. cerevisiae. Yeasts are unicellular organisms that reproduce asexually by budding. In general, yeasts are larger than most bacteria. Yeasts play an important role in the food industry as they produce enzymes that favour desirable chemical reactions such as the leavening of bread and the production of alcohol and invert sugar.

### 5.2.3 Moulds

Moulds are also important organisms in the food industry, both as spoilers and preservers of foods. Certain moulds produce undesirable toxins and contribute to the spoilage of foods. The Aspergillus species are often responsible for undesirable changes in foods. These moulds are frequently found in foods and can tolerate high concentrations of salt and sugar. However, others impart characteristic flavours to foods and others produce enzymes, such as amylase for bread making. Moulds from the genus Penicillium are associated with the ripening and flavour of cheeses. Moulds are aerobic and therefore require oxygen for growth. They also have the greatest array of enzymes, and can colonise and grow on most types of food. Moulds do not play a significant role in the desirable fermentation of fruit and vegetable products.

When micro-organisms metabolise and grow they release by-products. In food fermentations the by-products play a beneficial role in preserving and changing the texture and flavour of the food substrate. For example, acetic acid is the by-product of the fermentations of some fruits. This acid not only affects the flavour of the final product, but more importantly has a preservative effect on the food. For food fermentations, micro-organisms are often classified according to these by-products. The fermentation of milk to yoghurt involves a specific group of bacteria called the lactic acid bacteria (Lactobacillusspecies). This is a general name attributed to those bacteria which produce lactic acid as they grow. Acidic foods are less susceptible to spoilage than neutral or alkaline foods and hence the acid helps to preserve the product. Fermentations also result in a change in texture. In the case of milk, the acid causes the precipitation of milk protein to a solid curd.

### 5.2.4 The Starter Bacteria

In fermentation, the raw materials are converted by microorganisms (bacteria, yeasts and moulds) to products that have acceptable food qualities. Spontaneous (natural) fermentations, i.e., processes initiated without the use of a starter inoculum, have been applied to food preservation for millennia. In a natural fermentation, the conditions are set so that the desirable microorganisms grow preferentially and produce metabolic by-products, which give the unique characteristics of the product. The majority of small-scale fermentation in developing countries and even some industrial processes such as sauerkraut fermentation are still conducted as spontaneous processes. Various types of starter cultures and even back-slopping are widely used in fermentation processes, even in industralised countries. However, spontaneous food fermentations are neither predictable nor controllable. The natural microflora of the raw material is either inefficient, uncontrollable, and unpredictable, or is destroyed altogether by the heat treatments given to the food.

When the yield is unstable and where the desired microorganism might not grow, or where pathogenic microorganisms might also grow, a controlled fermentation is used. In a controlled fermentation, the fermentative microorganisms are isolated and characterised, then maintained for use. Since the addition of selected bacteria (usually LAB) starts the manufacturing process, they are commonly referred to as starter cultures. Starter cultures are added to the raw materials in large numbers and incubated under optimal conditions. In common controlled-fermented products such as sauerkraut and yoghurt, the primary function of LAB starters is the production of lactic acid from lactose which helps to make the products shelf-stable. Other functions of starter cultures may include flavour, aroma, and alcohol production, proteolytic and lipolytic activities, and inhibition of undesirable organisms. Therefore, microbial starters provide particular characteristics through fermentation in a more controlled and predictable way.

**There are two groups of commercial LAB starter cultures:**

- Defined (either single or multiple strains): composed of a known number of strains;
- Undefined (mixed): in which the number of strains is unknown.

Modern starter cultures are selected either as single or multiple strains, specifically for their adaptation to a substrate or raw material. Suitable cultures for fermentation must be selected at the strain level since not all strains of a species are equally suitable for use as starters, nor are all equally well adapted to a food substrate. In dairy products, the properties of starter cultures such as phage sensitivity, lactic acid, gas, and aroma compounds production are, to a large extent, dependent upon the properties of each culture componen. Because modern dairy starter cultures are blended empirically for the desired characteristics of the final product, maintenance of the optimal strain balance throughout the cheese fermentation process is important because the complex relationships among microorganisms can be easily be altered. Also small variations in microbial composition could have unexpected effects on cheese quality.

The largest part of the lactic culture market consists of cultures for the dairy industry. The size of this market is currently approximately US $250 million. The starter culture industry has been in a phase of consolidation and restructuring; in this context, commercial starter cultures for direct inoculation are increasingly used for their reliability, performance and safety, as well as for their convenience.

### 5.2.5 The Non-Starter Bacteria

In some fermented dairy products, additional bacteria such as Cit+Lactococcus lactis subsp. lactis, Leuconostoc spp., Lactobacillus kefir, and Propionibacterium freudenreichii subsp. germanii, often referred to as secondary microflora, are often intentionally introduced to produce aroma compounds and carbon dioxide in cultured buttermilk and certain cheeses. Other types of secondary microflora include undefined mixtures of yeasts (Debaryomyces spp. and Geotrichum spp.), moulds (Penicillium camemberti and Penicillium roqueforti), and bacteria (Brevibacterium linens, Micrococcus spp., and mesophilic lactobacilli). The use of micrococci and B. linens is usually limited to surface-ripened and mould-ripened cheeses; mesophilic lactobacilli, which can form biofilms and be a source of contamination in a dairy, are believed to contribute considerably to the formation of cheese aroma from amino acids. For example, a useful cooperation between starter L. lactis and glutamate – dehydrogenase-positive L. casei/L. paracasei strains to stimulate flavour development in Gouda and Cheddar cheeses has been suggested.

### 5.2.6 The Unwanted Bacteria

Food manufacturing, distribution, and storage rely on well-placed deleterious stresses or hurdles that either inhibit or inactivate contaminating microorganisms in food systems. However, fermented foods are not commercially produced in an environment free of contaminating microorganisms. For example, dairy products may contain yeasts, moulds, and many different genera of bacteria whose metabolic activities destroy quality. In cheese, after fermentation is complete, only acid-tolerant bacteria can grow. However, if acid development is slow or if the pH does not decrease sufficiently, contaminants which otherwise would have been inhibited may be able to grow. In some fermented foods the pH can increase during ripening and permits the growth of previously inhibited bacteria. These phenomena, together with the well-known evidences that many food-borne pathogenic bacteria are adapting to sub-lethal inimical stresses, may allow unwanted organisms (either pathogenic or spoilage) to grow in fermented foods.

## 5.3 DAIRY FERMENTATIONS

Two General Types:

**A. Cheeses**- Cottage, Cheddar, Italian, Swiss, etc.

1. Need acid-producing mesophilic or thermophilic bacteria depending on cook temp.
2. May include 2o fermentation by other microorganisms such as molds (blue or camembert), brevibacteria (limburger), or propionibacteria (Swiss). These may be added with the lactic starter or later in the production process.
3. Some varieties also include starters for flavor; e.g. L. lactis ssp. diacetylactis in Gouda, adjuncts in lowfat Cheddar.

**B. Semi-solid Milks**-Yogurt, sour cream, buttermilk, kefir

1. MO needed for acid and sometimes flavor production
2. Ethanol production in some (heterofermentive lactics and yeasts)
3. Source of beneficial bacteria (Lactobacillus acidophilus and Bifidobacterium).

### 5.3.1 Types of Cheese (Manufacturing Procedures)

**1. Cheddar Cheese**-most made from pasteurized milk, starter and rennet are added.

***Rennet Sources***

1. Calf stomach extract (traditional)
2. Microbial rennets (not inactivated by cook temps, can create bitterness in cheese)
3. Recombinant chymosin (a major share of the market)

***Key Steps for the Microbiology***

**a. Cook:** The temperature used to cook cheese essentially determines whether a thermophilic or mesophilic lactic acid bacteria will be used. Cheddar is cooked at 3940°C. Starter - mesophilic lactococci (L. lactis ssp. cremoris and L. lactis ssp. lactis)

**b. Cheddaring:** manufacturing step unique to Cheddar. During Cheddaring, the curd is spread evenly over the bottom of the vat after whey removal and allowed to knit for 15 min. A knife is used to cut the curd longitudinally down the middle, and horizontally at 10 inch intervals. The curd blocks are spaced about 1 inch apart, allowed to rest 15 min and then flipped over until the titratable acidity of the draining whey reaches 0.5-0.6% and then they are milled.

**c. Salting.** Salt is directly to the curd. Cheddar-salt is added at a level of 2-2.3% (w/w).

Salt adds flavor

Inhibits undesirable MO

Slows acid development from starters

Does not inhibit lactobacilli that dominate the 2o microbiota during ripening

**d. Ripening.** Cheddar is ripened at 6-10°C for 3 mo to a year (mild vs sharp). Starter proteolysis is important, secondary microbiota appear (lactobacilli, pediococci and micrococci) that provide proteases and lipases.

**2. Swiss Cheese:** Characterized by a sweet nutty flavor with eyes (gas pockets) Milk is not fully pasteurized to provide proper elasticity for eye development. Instead it is either heat treated at a sub-pasteurization temp (64-70°C/16 s) or it is treated with 0.05% $H_2O_2$. Excess peroxide is removed by adding catalase.

**a. Cook.** Cook temperature is 50-54oC, so a thermophilic starter consisting of S. thermophilus, L. delbuekii ssp. bulgaricus (or L. helveticus), and Propionibacterium freundenreichii is used. P. freundenreichii main role is to ferment lactate to carbon dioxide for eyes, propionic acid, and acetic acid. Swiss curd is pressed and hooped into wheels without cheddaring or milling.

**b. Salt**-Swiss cheese is brined in a saturated NaCl (23%) for 12-48 h at 8-12°C.

After brining, the cheese is stored dry at 7-13°C for 5-10 days to allow the salt conc. to equilibrate and to encourage the growth of propionibacteria. The salt concentration in Swiss is lower than Cheddar and can be low enough to permit growth of gaseous spoilage bacteria such as clostridia.

**c. Warm Room**-the cheese is then moved to a warm room (22-24°C) for 2-7 weeks where the propionibacteria grow and produce the eyes and the flavor compounds. Good eyes are smooth, 0.4-1.2 inches in diameter, round, and uniformly distributed.

Defects in Swiss cheese

**1. Blind**-no carbon dioxide produced so no holes or foamy Swiss

**2. Slit open eye defect**-too much gas produced or the curd is too brittle.

After the warm room, the cheese is stored at 1.5-4.5°C for 3-9 m for final ripening.

**3. Italian cheeses**

**a. Parmesan or Romano** have low moisture contents.

Made from 2.5% fat milk (pasteurized). The curd is cooked at 47-50°C so thermophilic starters are used.

Characterized by sharp volatile flavors which are the result of milk fat lipolysis. LAB do not have pronounced lipolytic activity, so lipase enzyme powders are added to the milk with the starter to generate the flavor notes.

A long brine period (2 weeks) is used to help reduce the moisture level. Parmesan and Romano may be aged at 4-15°C anywhere from 5-6 mo up to 1-2 yr.

**b. Pasta filata** (pulled curd) (Mozzarella) Most U.S. Mozzarella is used for pizza, and the annual Mozz production level is now equal to Cheddar cheese.

1. Mozzarella is cooked for 70 min at 42°C, so makers must use thermophilic starters (S. thermophilus and L. delbruekii subsp. bulgaricus or L. helveticus). Interestingly, it is also Cheddared but at the higher cook temperature
2. The cheddared curd is milled then mechanically mixed and heated in a device called a stretcher-cooker which blends the curd with 70°C water. This step is required to generate the elastic body that characterizes Mozzarella cheese.
3. The cheese is cooled, brined 2-4 h, then aged for a short time (7-30 days). Functional characteristics such as stretch, melt, cook color, and "oiling off" are critical properties.
4. Surface ripened cheese. (Muenster, limburger)

a. Limburger is not salted. It is simply hooped and stored at 15-16oC for 4-10 days at 95% RH on wooden racks that provide the surface inoculum of yeasts and Brevibacterium linens required for proper flavor development. These MO grow on the outer surface of the cheese.

The yeasts metabolize lactate which raises the pH on the curd surface to about 5.9 which allows B. linens to grow. Within a few days a characteristic reddish-brown slime of B. linensappears. Proteolytic and lipolytic enzymes of B. linens then diffuse into the cheese and produce the characteristic flavor and aroma of Limburger.

Limburger is kept in the humid room for 20 days, wrapped (without removing the surface smear), and then stored at 4.5°C for 21 days.

b. Muenster is brined for 2 days at 10°C, and then surface ripening is allowed to proceed for 2-4 days. The smear is dried and annatto is wiped on to give the cheese an orange exterior.

**a. Blue veined cheese: Blue or Roquefort.**

"Blue cheese"-made from cows milk

"Roquefort"- True Roquefort cheese is made only in the Causses region of southern France, from full cream milk collected from a special breed of sheep (8% fat vs 3.5% for most cows) and ripened in the famous Roquefort caves.

Because cows milk cream is not as white as sheeps, milk for blue cheese in the U.S. is separated, and the cream may be bleached with benzoyl peroxide to obtain the whiteness desired in this variety. The milk is pasteurized, then recombined and homogenized (increases the surface area of fat available for mold lipolytic enzymes).

The curd is cooked at 30-33°C so our acid producer will be mesophilic lactococci.

Curd is cut into larger cubes to help create the open body texture required for proper mold growth. After the whey is drained, 2% salt and 0.0625% powdered spores (1 oz/100 lb curd) of Penicillium roqueforti are sprinkled over the curd. The curd is hooped, pressed, and allowed to drain 12-18 h.

The cheese is salted again the next day by rubbing dry salt on the surface.

After salting, the cheese block is pierced with a gang of needles to facilitate gas exchange for mold growth. This step creates the "veins" for mold growth. Mold grows during subsequent storage at 10-13°C in 95% RH and requires 2-4 mo. The mold produces lipases that hydrolyze milk fat to release free fatty acids and their breakdown products. These compounds create the sharp peppery flavor and characteristic aroma of Blue cheese.

**Cheese with surface mold (Camembert, Brie).** In most countries the only difference between these two cheeses is the size of the cheese wheel, but in France, true Brie cheese has surface mold and a reddish smear of B. linens on its surface.

Made from pasteurized milk, the curd is not cooked-just kept at 32.2°C from start to finish, so mesophilic lactococcal starters are used.

Penicillium camemberti spores are sprayed onto the cheese and cheese is given a 30 minute brine bath. The absence of a cooking step and use of only light salting results in a soft, high moisture cheese.

To promote mold growth, the cheese is stored at 14°C for 1-2 weeks at 95% RH. A white mat of mold develops, and the cheese is turned on the racks to obtain uniform growth. Mold lipases and proteases penetrate the cheese to produce the flavor.

Mold covered wheels are wrapped in foil and ripened at 5°C for 2-5 weeks. Cheeses must be eaten within 6-7 weeks of manufacture or the flavors become too strong.

## 5.4 IMPACT OF LAB ON CHEESE FLAVOR

1. Fermentation and depletion of fermentable sugars
2. Creating a low oxidation-reduction potential. This is thought to promote the nonenzymatic evolution and stabilization of volatile flavor compounds (e.g. sulfur-containing cmpds in Cheddar)
3. Synergism among bacteria. Lactate produced by the starter during manufacture and ripening can serve as substrate for 2o bacteria
4. Lipolysis.
5. Proteolysis
6. Synthesis of cheese flavor compounds. Although LAB are thought to catalyze synthesis of other flavor compounds, most of these compounds remain unidentified.

### 5.4.1 Fermented Milks

Absence of a whey removal step differentiates these products from cheese.

**A. Yogurt:** To obtain the characteristic body of yogurt, milk is concentrated. Gums or stabilizers may be added to reduce syneresis and to obtain a desired texture. Alternatively, starters that produce an extracellular polysaccharide slime are used to obtain similar results.

Yogurt milk is homogenized, then heat treated at 80-85°C for 30 min.

The milk is cooled to 42oC, inoculated, then incubated at 40-45°C for 8-24 h, then cooled and sold quickly. The high incubation temperature requires use of thermophilic cultures. The two starters exhibit symbiotic growth in milk.

1. L. delbruekii subsp. bulgaricus is more proteolytic than S. thermophilus so the lactobacilli degrade casein and provide S. thermophilus with the essential amino acids (valine, leucine, histidine, and isoleucine) needed for rapid growth.
2. S. thermophilus produces trace amounts of formic acid and $CO_2$ which stimulate the growth of L. delbruekii subsp. bulgaricus.

The net result is that growth of the organisms together produces more lactic acid than either culture is capable of producing alone.

The most important flavor compound in yogurt is acetaldehyde (2.4-41 ppm).

Final product contains 108 to 109 CFU/g but counts decrease with storage time. Normal starter strains are unable to colonize the gut.

**Two Main Types of Yogurt**

**a. Set:** The product formed when fermentation and coagulation of milk is carried out in the retail container, and the yogurt produced is in a continuous semi-solid mass.

**b. Stirred:** Results when the coagulum is produced in bulk and the gel structure is broken before cooling and packaging.

**B. "Therapeutic" Milks:** Consumed primarily for the microorganisms (probiotics) they contain, which are perceived to be beneficial to human health.

Some studies have suggested that a variety of health and nutritional benefits may arise from ingestion of L. acidophilus, a few other lactic acid bacteria, and bacteria in the genus Bifidobacterium. Many are normal components of the human intestinal tract. The conclusions of studies have been contradictory and most health claims remain controversial.

**1. Acidophilus Milk:** Milk containing a pure culture of L. acidophilus. Traditional acidophilus milk is simply a sour milk product. To get a product which was more acceptable to the American palate, yet still contained high numbers of the therapeutic organism, an unfermented acidophilus milk product called sweet acidophilus milk was developed. Milk is inoculated with a high number of L. acidophilus harvested after growth in another medium. Sweet acidophilus milk contains over 500 million CFU/ ml and is more accepted by the American public.

**2. Acidophilus and Bifidus Yogurt:** Use L. acidophilus as the lactobacillus component, bifids sometimes included.

**C. American Buttermilk:** Uses both lactic acid producers and citrate fermenters (Lactococci, Leuconostoc). Flavor compounds are lactic acid, diacetyl, and acetoin. Most flavor compounds excluding lactate are formed from the utilization of citrate. In this product, Pasteurized milk is inoculated and incubated at 21-24°C until a coagulum is formed, usually 14-16 hrs. The product is then cooled and then agitated to break up the coagulum.

Milk is usually fortified with Na-citrate at a rate of 2-2.5 lbs per 100 gallons to ensure substrate will be available to produce diacetyl. Milk contains.14 to.17% citrate.

**D. Alcoholic Milks**

1. **Kefir (Asia)**-uses kefir grains: rice-sized mass of MO which include:

   Lactobacillus, Leuconostoc, lactococci and 2 yeasts bound together by a polysaccharide coat (looks like grain of rice) Manufacture: pasteurized milk inoculated with kefir grains at 22OC, soured overnight.

   Kefir grains filtered out and used again.

   Final product has a cheese-like smell and taste, viscous, slimy appearance and feel.

   0.8% lactic acid,1% EtOH, saturated w/ $CO_2$ so it foams like beer.

2. **Koumiss**

   Traditionally made from mares milk, modern bulk production in CIS uses pasteurized skimmed cows milk.

   Fermented by Lactobacillus delbruekii ssp. bulgaricus with Saccharomyces and Candida sp. yeasts final product has: 0.7-1.8% lactic acid, 1-2.5% EtOH

   Used therapeutically by people in Russia for pulmonary tuberculosis.

## 5.5 STARTER CULTURES

The preservation of food by fermentation is one of the oldest methods known to mankind. A typical example is lactic acid fermentation, which is widely used for the preparation of several fermented milk products, such as dahi (curd), yoghurt, acidophilus milk, shrikhand and various varieties of cheeses. In the modern dairy industry, dairy starter cultures are pre-requisite for the production of safe products of uniform quality. Lactic acid bacteria are often called dairy starter cultures, which are used for the production of various fermented milk products.

Dairy starter cultures are carefully selected microorganisms, which are deliberately added to milk to initiate and carry out desired fermentation under controlled conditions in the production of fermented milk products. Most of them belong to lactic acid bacteria (Lactococcus, Lactobacillus, Streptococcus and Leuconostocs). In some cases, few non-lactic starters (bacteria, yeast and mold) are also used along with lactic acid bacteria during manufacturing of specific fermented milk products, such as kefir, kumiss and mold ripened cheeses.

Starter cultures can be used as single strain, mixed strain and multiple strains depending upon the type of products to be prepared. The ability of starter culture to perform its functions efficiently during manufacture of fermented dairy foods depends primarily on purity and activity of starter cultures.

Starter cultures are those microorganisms that are used in the production of cultured dairy products such as yogurt and cheese. The natural microflora of the milk is either inefficient, uncontrollable, and unpredictable, or is destroyed altogether by the heat treatments given to the milk. A starter culture can provide particular characteristics in a more controlled and predictable fermentation. The primary function of lactic starters is the production of lactic acid from lactose. Other functions of starter cultures may include the following:

- Flavour, aroma, and alcohol production
- Proteolytic and lipolytic activities
- Inhibition of undesirable organisms

There are two groups of lactic starter cultures:

1. Simple or defined: single strain, or more than one in which the number is known
2. Mixed or compound: more than one strain each providing its own specific characteristics

Starter cultures may be categorized as mesophilic, for example:

- Lactococcus lactis subsp. cremoris
- L. delbrueckii subsp. lactis
- L. lactis subsp. lactis biovar diacetylactis
- Leuconostoc mesenteroides subsp. cremoris or thermophilic:
- Streptococcus salivarius subsp. thermophilus (S.thermophilus)
- Lactobacillus delbrueckii subsp. bulgaricus
- L. delbrueckii subsp. lactis
- L. casei
- L. helveticus
- L. plantarum

Mixtures of mesophilic and thermophilic microorganisms can also be used as in the production of some cheeses.

### 5.5.1 Types of Starter Culture

Several authors have produced definitions of starters including Lawrence et al. (1976). The latter has classified the three main types of starter used commercially in Australia, New Zealand, the UK and North America as follows:

a. Single-strain starters: single strains of Str. cremoris and less commonly Str. lactis. These have been used in pairs in some factories in New Zealand and in Scotland but also singly in Australia.

b. Multi-strain starters: defined mixtures of three or more single strains of Str. cremoris and/or Str. lactis. Leuconostoc and Str. diacetylactis strains may also be used. Multiple-strain starters are frequently referred to as mixed-strain starters in the United States of America.

c. Mixed-strain starters: mixtures of strains of Str. cremoris, Str. lactis, Str. diacetylactis and leuconostocs. The identity of the component strains is frequently unknown to the user and their composition may vary on subculture.

This classification is now of limited value for several reasons. This scheme was developed before the combined use of lactococci and Str. thermophilus as starters for Cheddar and similar cheeses became common. Incidentally the so called stabilised cultures, used in the manufacture of some soft cheeses contain blends of lactococci and Str. thermophilus. In practice it now usual to define starters as either defined, meaning that the strain and species of the component strains are known or as undefined cultures. Included in this latter grouping are the artisanal cultures widely used in some European countries, particularly in Italy.

## 5.6 PROBIOTICS

Probiotics are defined as live microorganisms that, when administered in adequate amounts, confer a health benefit on the host. Probiotic organisms used in foods must have the ability to resist gastric juices, exposure to bile, and be able to proliferate and colonize the digestive tract. The beneficial effects of probiotic foods on human health and nutrition are constantly increasing, and probiotics are popularly using bio-ingredients in many functional fermented foods. The most commonly used probiotic bacteria belong to the heterogeneous group of LAB (Lactobacillus, Enterococcus) and to the genus Bifidobacterium, however, yeasts and other microbes have also been developed as potential probiotics during recent years. Some popular commercial probiotic cultures which are available in global markets include Bacillus coagulans BC30 marketed by Ganeden Biotech, Inc., Cleveland, OH, USA; Lactobacillus acidophilus NCFM, Lactobacillus rhamnosus HN001 (DR20) and Bifidobacterium lactis HN019 (DR10) marketed by Danisco (Madison, WI, USA), L. casei strain Shirota and B. breve strain Yakult marketed by Yakult (Tokyo, Japan), L. fermentum VRI003 (PCC) marketed by Probiomics (Eveleigh, NSW, Australia), L. rhamnosus R0011 marketed by Institut Rosell (Montreal, QC, Canada), Streptococcus oralis KJ3 marketed by Oragenics, Inc. (Alachua, FL, USA), and Saccharomyces cerevisiae (boulardii) marketed by Biocodex.

Products containing probiotic bacteria generally include foods and supplements. Fermented milk products are the most traditional source of probiotic strains of lactobacilli; however, commercial probiotic lactobacilli have also been added to meat products, snacks, fruit juice, etc. Probiotic properties of Lactobacillus plantarum isolated from kimchi, Korean fermented vegetable product, has been reported, and is also found to prevent the growth of Helicobacter pylori. Probiotic strain L. acidophilus La-5 produces conjugated linoleic acid (CLA), an anti-carcinogenic agent. Pediococcus pentosaceus CIAL-86 isolated from wine shows anti-adhesion activity against Escherichia coli CIAL-153, indicating its probiotic potential in wine.

The internationally endorsed definition of probiotics is live microorganisms that, when administered in adequate amounts, confer a health benefit on the host. Other definitions advanced through the years have been restrictive by specification of mechanisms, site of action, delivery format, method, or host. Probiotics have been shown to exert a wide range of effects. The mechanism of action of probiotics (e.g., having an impact on the intestinal microbiota or enhancing immune function) was dropped from the definition to encompass health effects due to novel mechanisms and to allow application of the term before the mechanism is confirmed. Physiologic benefits have been attributed to dead microorganisms. Furthermore, certain mechanisms of action (such as delivery of certain enzymes to the intestine) may not require live cells. However, regardless of functionality, dead microbes are not probiotics.

The term "probiotic" is sometimes erroneously used as a synonym for putatively beneficial members of commensal microbiota. The context for this misuse is the assertion that certain dietary or environmental factors may "encourage your native probiotics." Members of human commensal microbiota are often sources from which probiotics are isolated, but, until such strains are isolated and then adequately characterized for content, stability, and health effects, they are not probiotics.

The US Food and Drug Administration (FDA) uses other terms for live microbes for regulatory purposes; live microbes used in animal feeds are called "direct-fed microbials", and, when intended for use as human drugs, they are classified as "live biotherapeutics". However, no legal definition of probiotics exists in the United States or in other countries, which allows the marketing of products labeled as "probiotics" that do not meet the fundamental criteria stipulated in the scientific definition. As public awareness of probiotics increases, the use of the term has implications both for the violation of the standard of truthful and not misleading labeling and for consumer confidence in this product category. In the absence of a legal definition of the term "probiotic," it is incumbent on industry participants to adopt and adhere to the scientific definition of the term. If the products are not properly characterized and validated, the category will suffer.

### 5.6.1 Commercially Important Fermentation

1. Microbial cells or Biomass as the product: Eg. Bakers Yeast, Lactic acid bacillus, Bacillus sp.
2. Microbial Enzymes: Catalase, Amylase, Protease, Pectinase, Glucose isomerase, Cellulase, Hemicellulase, Lipase, Lactase, Streptokinase etc.
3. Microbial metabolites :
   - o Primary metabolites – Ethanol, Citric acid, Glutamic acid, Lysine, Vitamins, Polysaccharides etc.
   - o Secondary metabolites: All antibiotic fermentation
   - o Recombinant products : Insulin, HBV, Interferon, GCSF, Streptokinase
   - o Biotransformations: Eg. Phenyl acetyl carbinol,Steroid Biotransformation

## 5.7 NUTRIENT SOURCES FOR INDUSTRIAL FERMENTATION

### 5.7.1 Nutrient Raw Material

***Carbon Source***

Glucose Corn sugar, Starch, Cellulose

Sucrose Sugarcane, Sugar beet molasses

Lactose Milk whey

Fats Vegetable oils

Hydrocarbons Petroleum fractions

***Nitrogen Source***

Protein Soybean meal, Cornsteep liquor, Distillers' solubles

Ammonia Pure ammonia or ammonium salts

Nitrate Nitrate salts

Nitrogen Air

***Phosphorous Source Phosphate Salts***

**Trace elements :** Fe, Zn, Cu, Mn, Mo, Co

**Antifoaming agents :** Esters, Fatty acids, Silicones, Sulphonates, Polypropylene

**Buffers:** Calcium carbonate, Phosphates

**Growth Factors:** Some microorganisms cannot synthesize the required cell components themselves and need to be supplemented: E.g. Thiamine, Biotin, Calcium pentothenate

**Precursors:** Directly incorporated into the desired product: Phenyl ethylamine into Benzyl penicillin, Phenyl acetic acid into Penicillin G

**Inhibitors:** To get the specific products: e.g. Sodium barbital for Rifamycin

**Inducers:** Majority of the enzymes are inducible and are synthesized in response of inducers: e.g. Starch for Amylases, Maltose for Pollulanase, Pectin for Pectinase

**Chelators:** Chelaters are the chemicals used to avoid the precipitation of metal ions. Chelaters like EDTA, Citric acid, Polyphosphates are used in low concentrations.

### 5.7.2 Biotechnology Companies Developing Pharmaceuticals via Fermentation in Transgenic Microbes or Mammalian Cell Cultures

- Atlantic BioPharmaceuticals, Inc.
- Chiron Corporation
- BioPharmaceuticals Pipeline
- Genentech, Inc.
- Interomex Biopharmaceuticals
- Lonza Biotechnology
- Marathon Biopharmaceuticals, Inc.
- NPS Pharmaceuticals
- Protein Design Labs
- Protein Sciences Corp

## 5.8 PRODUCTS OF FERMENTATION PROCESSES

The growth of micro-organisms or other cells results in a wide range of products. Each culture operation has one or few set objectives. The process has to be monitored carefully and continuously, to maintain the precise conditions needed and recover optimum levels of products. Accordingly, fermentation processes aim at one or more of the following:

(a) Production of cells (biomass) such as yeasts;

(b) Extraction of metabolic products such amino acids, proteins (including enzymes), vitamins, alcohol, etc., for human and/or animal consumption or industrial use such as fertiliser production;

(c) Modification of compounds (through the mediation of elicitors or through biotransformation); and

(d) Production of recombinant products.

**A. Microbial biomass:** Microbial biomass is produced commercially as single cell protein(SCP) using such unicellular algae as species of Chlorella or Spirulinafor human or animal consumption, or viable yeast cells needed for the baking industry, which was also used as human feed at one time. Bacterial biomass is used as animal feed. The biomass of Fusariumgraminearum is also produced, for a similar use.

**B. Microbial metabolites (Primary metabolites):** During the log or exponential phase organisms produce a variety of substances that are essential for their growth, such as nucleotides, nucleic acids, amino acids, proteins, carbohydrates, lipids, etc., or by-products of energy yielding metabolism such as ethanol, acetone, butanol, etc. This phase is described as the tropophase, and the products are usually called primary metabolites.

**C. Food industru products:** A very wide range of innumerable products of the food industry, such as sour cream, yoghurt, cheeses, fermented meats, bread and other bakery products, alcoholic beverages, vinegar, fermented vegetables and pickles, etc., are produced through microbial fermentation processes. The efficiency of the strains of the organisms used, and the processes are being continuously improved to market quality products at more reasonable costs.

**D. Recombinant Products:**: Recombinant DNA technology has made it possible to introduce genes from any organism into micro-organisms and vice versa, resulting in transgenic organisms and the latter are made to produce the gene product. Genetically manipulated Escherichia coli, Saccharomyces cerevisiae, other yeasts and even filamentous fungi are now being used to produce interferon, insulin, human serum albumin, and several other products.

**E. Biotransformation:** Production of a structurally similar compound from a particular one, during the fermentation process is transformation, orbiotransformation, or bioconversion. The oldest instance of this process is the production of acetic acid from ethanol.

Immobilised plant cells may be used for biotransformation. Using alginate as the immobilising polymer, digitoxin from Digitalis lanatawas converted into digoxin, which is a therapeutic agent in great demand. Similarly, codeinone was converted into codeine and tyrosine from Mucuna pruriens was converted into DOPA.

**F. Elicitors:** It is possible to induce production or enhance production of a compound in cultures by using elicitors, which may be micro-organisms. For example, Saccharomyces cerevisiae was an efficient elicitor in the production of glyceollin (Glycine max) and berberine (Thalictrum rugosum). Rhizopus arrhizus trebled diosgenin production by Dioscorea deltoidea. The production of morphine and codeine byPapaver somniferum was increased 18 times by Verticillium dahliae

## 5.9 MICROORGANISMS IN FOOD ECOSYSTEMS

The importance of ecological concepts in understanding the growth of microorganisms in foods is well recognised by food microbiologists. These ecological principles are the fundamental to modern quality assurance, predictive modelling, and risk analysis strategies to prevent outbreaks of food spoilage and food-borne disease. They also form the basis for the use of microorganisms in the production of fermented foods and beverages, and for their use as probiotic, starter, and biocontrol agents.

### 5.9.1 Ecological and Biochemical Factors Affecting Microbial Growth in Food

According to Fleet, to effectively manage the growth and activities of microorganisms in fermented foods, the following points should be raised:

- Information on diversity, taxonomic identity, growth cycle, quantitative changes, and spatial distribution of microbial species that ferment into the food at every stage of production;
- Biochemical and physiological data on the food colonisation process;
- Impact of the intrinsic, extrinsic, and processing factors influencing microbial growth, survival and biochemical activity;
- Relationship between growth and activity of individual microorganisms and product quality and safety.

### 5.9.2 Problems Related to Qualitative and Quantitative Estimation of Microbial Populations in Foods

One of the most important applications of ecological concepts to food fermentation ecosystems is the identification, interpretation and manipulation of time-dependent changes in communities of microorganisms. The economic and social consequences of microorganisms in foods depend not only on the species present but, most importantly, on their numbers. The number of microbial cells ultimately determines whether or not the product will cause an outbreak of disease or develop an off-flavour. Unfortunately, the vast majority of ecological studies in food microbiology fall significantly short of providing this quantitative knowledge. Many studies simply describe the isolation and identification of the "most predominant" species at one point in the product's history, while others have provided semiquantitative data by reporting the frequency of isolation of specific organisms. Population changes are mostly described in reference to microbial groups (e.g., total plate count, coliforms, LAB) rather than data for particular species or strains, which could be more useful to better comprehend the final microbiological properties of fermented foods.

In a recent and excellent review article, limitations of the methodology to recover microbes from foods, particularly concerning erroneous data given by maceration and dilution steps and quantitative errors coming from underestimation of uncultivable cells, have been summarised. Furthermore, communities of any organisms are static neither in space nor time. This phenomenon, which in ecology is called spatial and temporal heterogeneity, is determined by a sequential development of species and strains occurring in many products, which leads to a continuously changing microbial content, both qualitatively and quantitatively. Heterogeneity results from inherent variability in organic structure and composition of the food (e.g., in surface-smeared cheeses), from changes in the external climatic environment, and as a result of microbial activity.

The dynamics of growth, survival and biochemical activity of microorganisms in foods are the result of stress reactions in response to the changing of the physical and chemical conditions into the food microenvironment (e.g., the gradients of pH, oxygen, water activity, salt, and temperature) and the ability to colonise the food matrix and to grow with spatial heterogeneity (e.g., microcolonies and bio-films). Moreover, because in most food ecosystems microbial populations are assembled to give a community, and because cells are generally immobilised and localised in high densities, food production (or degradation) is rarely the result of the activities of an individual but that of a group of microorganisms. Therefore, the growth, survival, and activity of any one species or strain, whether it be an unwanted spoilage or pathogenic organism, or a desirable biocontrol of a probiotic agent, will, in most cases, be determined by the presence of other microorganisms and the in situ cell-to-cell ecological interactions which often happen in a solid phase. It has been shown that in situ solid phase associations similar to those found in food ecosystems could induce unique biochemical and physiological reactions of a microbial population as a whole.

Understanding the patterns of occurrence of microbial populations in food is really a problem of determining the actual niche of the organism present; that is, identifying the location and the role of an organism in both space and time. Reliable quantitative ecological data should, therefore, take into consideration the dynamics of microorganisms in food ecosystems. This information is of key importance in food ecology, especially to understanding the behaviour of pathogens in foods.

### 5.9.3 Survival Mechanisms of Microorganisms in Foods

Microbial ecology is microbial physiology under the worst possible conditions. It is widely accepted that plate culturing techniques reveal a small portion (i.e., that viable and culturable) of the true microbial population in natural ecosystems, and that this can essentially be explained by two ecological factors: (i) an inability to detect novel microorganisms, which might not be cultivable with known media and (ii) an inability to recover known microorganisms which are either stressed or actively growing but which enter a non-cultivable state.

### 5.9.4 The Sub-Lethally Injured Bacteria

Stress is any change in the genome, proteome or environment that imposes either reduced growth or survival potential. Such changes lead to attempts by a cell to restore a pattern of metabolism that either fits it for survival or for faster growth. For any stress, the bacterial cell has a defined range within which the rate of increase of colony forming units (CFUs) is positive (growth), zero (survival) or negative (death). In the first two cases, i.e., growth and survival, the cells are sub-lethally injured, whereas in the case of death (e.g., after a bacteriophage attack) the cells are lethally damaged and rapidly autolyse. The individual values at which the cell moves from one physiological state into the next is conditional on the degree of stress imposed by other environmental conditions.

In microbial populations, viable cells are usually countable on both non-selective and selective media, whereas stressed cells are able to form colonies on non-selective media but are not countable on selective media. In foods, many adverse conditions such as nutrient depletion, low temperature and other stresses can sub-lethally damage microorganisms. The viable but non-cultivable (VNC) state is a kind of stress which induces healthy, cultivable cells to enter a phase in which they are still capable of metabolic activity but do not produce colonies on media (both non-selective and selective) that normally support their growth. Sub-lethally injured cells (including VNC cells) are, however, usually able to resume their healthy state.

The VNC state has been shown in both Gram positive and Gram negative bacteria in the natural environment and it has also been experimentally induced in most food-borne pathogens. Although no molecular technologies have been systematically applied to food ecology, it is conceivable that food-associated microorganisms might enter the VNC state under certain situations, e.g., on the rind of surface-ripened cheese or on the surfaces of fruits and vegetables, where nutrient is limiting. It has been recently suggested that LAB can survive in a VNC state during wine storage and ageing.

## 5.10 MONITORING BACTERIAL DYNAMICS IN FOOD

### 5.10.1 Culture-Dependent Methods

They include traditional cultivation methods in combination with phenotypic (physiological and biochemical) and genotypic (species-specific and RAPD-PCR) identification and typing techniques of isolated strains. Generally, microbiota present during fermentation is enumerated by cultivating on media with the intention of selecting different groups of bacteria (e.g., total aerobic count on PCA medium, presumptive streptococcal and lactococcal counts on LM17 and LM17 plus cycloheximide, respectively, enterococci on KAA medium, leuconostocs on MSE medium, mesophilic lactobacillus on FH medium, thermophilic lactobacillus on MRS medium, etc.). Differences in oxygen tolerance, nutritional requirements, antibiotic susceptibility, and colony morphology and colour constitute the bases of differentiation of these methods. Selectivity of enumeration, however, can be strongly improved by altering incubation temperatures, incubation times, and the pH of the medium. Media allowing the separate (or simultaneous) enumeration of Lactobacillus delbrueckii subsp. bulgaricus, Streptococcus thermophilus, L. acidophilus, and Bifidobacterium species (such as B. bifidum and B. adolescentis) are available today for quality control of bioyoghurts.

Classical plating methodologies, therefore, only allow a rough measure of microbial groups or genera, and rarely is their sensitivity down to the species level. As stated above, the isolation, identification, and (facultative) typing of many strains from agar plates are needed to increase sensitivity. For the latter two purposes, either phenotyping or genotyping (or both) is applied. Concerning phenotyping, identification down to species level can be performed by API galleries or other phenotypic methods such as the Phene PlateTM system. These techniques allow dynamic trends of cultivable, dominant microbial populations or microbial groups to be highlighted. In recent years, however, genotyping (especially RAPD-PCR) has been successfully applied to identify the presence, succession, and persistence of LAB starter cultures inoculated in silage, sour doughs, probiotic cheese, and meats, as well as to determine the variations of microbial populations in naturally fermented products such as artisan cheeses, olives, meats, sour doughs, and indigenous fermented foods. A variation of this approach is the use of colony hybridisation using group-specific probes to allow specific differentiation and quantification of target population(s) without the need to subculture isolated colonies. Erlandson and Batt described a method using hydrophobic grid membrane filter colony hybridisation for quantitative strain-specific detection of lactococci in bacterial populations. However, classical plating methods to monitor microorganisms are laborious and time-consuming, especially when both spatial and temporal distribution of microbial populations are needed. Most importantly, culture-dependent methods are not effective if, for any reason, microorganisms are not cultivable.

### 5.10.2 Culture-Independent Methods

To determine the diversity of different microorganisms in natural ecosystems and to monitor evolution of microbial populations over space or time, culture-independent methods have been developed. Compared to traditional culturing, these methods (generally nucleic acids-based) are aimed at obtaining both a qualitative and quantitative picture of a microbial community without the need to isolate and culture its single components. Culture-independent methods have limited applications to food ecosystems.

### 5.10.3 Applications to Foods

Recent applications of DGGE (or TGGE) in identifying major rRNA species present in mixed populations allowed assessment of the impact of probiotic bacteria on the native human gastrointestinal microflora. DGGE of PCR-amplified products from the variable region of 16S rRNA genes has been recently used to identify Lactobacillus spp. isolated from naturally fermented Italian sausages, and to evaluate bacterial diversity occurring in different types of Mozzarella cheese. TGGE was applied to controlled dairy ecosystems with defined compositions, including liquid (starter), semisolid (home-made fermented milk), and solid (miniature cheese models) matrices, and to commercial dairy products. Subspecies, species, or groups of species of LAB were distinguished in dairy samples. LH-PCR was applied to depict population structure of the LAB species recoverable from Grana

Padano cheese whey starters. Typical bacterial species present in the LAB community were well discriminated. Small differences in species composition, e.g., the frequent finding of S. thermophilus and the constant presence of thermophilic lactobacilli (Lactobacillus helveticus, L. delbrueckii subsp. lactis/bulgaricus, and Lactobacillus fermentum) were reliably highlighted. When present, non-dominant bacterial species present in the whey starters, such as S. thermophilus and mesophilic lactobacilli, were easily visualised and characterised without isolating and cultivating single strains. Sensitivity of LH-PCR was confirmed by species-specific PCR from total DNA of the cultures.

By fluorescence-in situ-hybridisation (FISH) whole fixed cells can be identified and counted directly, i.e., in minimally disturbed samples. In situ identification of lactococci, enterococci, and streptococci allowed different microbial species in mixed cultures to be distinguished. Specific detection and enumeration of carnobacteria and Lactobacillus brevis from sea food products by using ISH flow cytometry has been recently performed. Rapid and sensitive ISH methods have been used for detecting and identifying LAB and brevibacteria in wine and on the surface of Gruyère cheese, respectively. In this latter work, it was demonstrated that only a minor part of the microorganisms resident on the cheese surface was cultivable. Millet and Lonvaud-Funel recently found the same behaviour in LAB community enumerated by colony counting and direct epifluorescence technique (DEFT) during wine storage. These latter two examples confirm that classical cultivation techniques are sometimes not able to reveal the most abundant microbes in a given ecosystem. Therefore, the term “dominant” should be used carefully when speaking of food-associated populations in the absence of in situ techniques for microbial counting.

Immunomagnetic separation has been mostly used in the separation of pathogenic bacteria from enrichment cultures and foods. FT-IR has been applied as an alternative to the traditional taxonomic techniques to discriminate and identify pathogens, spoilage contaminants, and LAB in foods.

It should be noted that most of the reported studies have focused on the analysis of end products. DGGE and other techniques have, however, more successfully been applied in polyphasic studies to monitor the microbial dynamics of food ecosystems. By combining different methods (e.g., PCR-DGGE, cloning and sequencing of rDNA amplicons and classical cultivation techniques) it has been possible to profile time-dependent specific changes in the composition of complex microflora, evaluate and quantify non-culturable microbial populations, and among these latter, to monitor the metabolically active microbial groups (or species-groups). This latter point is strongly related to the need to understand the role of unculturable microorganisms in a complex food ecosystem. This can be applied through culture-independent characterisation linked to determination of in situ metabolic activity. Even when an organism can be cultivated, properties determined in the laboratory may not necessary reflect the activities and physiology of their counterparts in the ecosystem.

## 5.11 CHEESE FERMENTATION

DGGE profiles derived from PCR and RT-PCR of DNA and rRNA, respectively, of milk and cheese samples of Ragusano cheese, an Italian artisan cheese produced in Sicily, were compared to determine the expression level of the 16S rRNA genes of the most predominant bacteria, which may reflect their contribution to the total activity. It was possible to: (i) qualitatively profile the dynamics of prevailing species: in the raw milk (Leuconostoc mesenteroides, Lactococcus lactis), during lactic fermentation (S. thermophilus), and during ripening (L. delbrueckii, L. fermentum); (ii) evaluate a substantially different degree of metabolic activity for the microbial groups detected, i.e., L. lactis and L. mesenteroides, both active in milk but not metabolically active in the ripened cheese, and L. delbrueckii subsp. bulgaricus, which showed an opposite trend; (iii) show a different composition of species identified after cultivation by phenotypic characterisation and those detected by PCR-DGGE, e.g., L. delbrueckii, which dominated in the ripening cheese but was not detected by cultivation. The use of RT-PCR is based on the evidence that active bacteria have generally higher numbers of ribosomes than dead or dormant cells. In other words, RT-PCR would provide a rough indication of the relative activities of the major bacteria present.

In practice, the low sensitivity of this technology (50 times lower than direct PCR) and the false positivity often detected from non-viable, mild-treated cells (whose ribosomal proteins would protect rRNA from endogenous RNAse degradation) underlines the usefulness of RT-PCR in gene expression studies rather than as a cell viability index. In this regard, the application of RT-PCR to messenger-RNA (mRNA) could be more successful because these molecules have a shorter half-life (40 s 20 min) than rRNAs, even in dead cells. By RT-PCR of mRNA it would also be possible to increase our knowledge on gene expression of phenotypic features, especially during food fermentation and manufacturing. In this way, not only the presence of an organism but also some specific metabolic activities of interest could be measured.

A polyphasic PCR-DGGE approach was used to describe the microbial population occurring in natural whey cultures for water-buffalo Mozzarella cheese production. By PCR-DGGE of the 16S rDNA V3 region, the total microbial community was assessed without cultivation by analysing DNA directly extracted from original whey samples and compared to the cultivable community assessed by analysing DNA extracted from combined colonies harvested from different agar plate media. In addition to the four common species evidenced by both total and cultivable community profile (i.e., S. thermophilus, L. lactis, L. delbrueckii, and L. crispatus), two additional species (i.e., L. fermentum and E. faecalis) were highlighted only by the cultivable community. The authors concluded that PCR amplification in the analysis of the total community DNA introduced bias.

## 5.12 FERMENTATION OF CEREAL PRODUCTS

Both the qualitative dynamics of the bacterial population and tracing the source of different strains dominating in sourdough fermentation processes using different starter mixtures were shown to be possible by PCR-based methods and plate culturing. The dynamics of both cultivable and non-cultivable bacteria were monitored by PCR-DGGE and sequencing of partial 16S rDNA genes. Depending on the prevailing conditions in the different sourdough fermentation, only a few Lactobacillus species were found to be competitive and became dominant; the results of PCR-DGGE were not always consistent with those obtained by culturing. Isolates of the species L. sanfranciscensis and L. fermentum were shown by RAPD-PCR analysis to originate from the commercial starters and the baker's yeast, respectively, used to produce the sourdoughs.

A polyphasic approach was applied to evaluate the dynamics of the microbial community responsible for the traditional fermentation of maize in the production of Mexican Pozol. The results confirmed that different culture-dependent or -independent techniques are necessary to obtain a more realistic view of the microbial ecology of this food. In particular, three stages in the microbial dynamics could be clearly detected. Similarly, an integrated strategy using culture-dependent and culture-independent methods was applied to study representative fermenting bacteria from natural cassava dough fermentation. In particular, some Lactobacillus species detected in dough samples by sequence analysis of DGGE bands were not recovered under any of the different culture media and conditions used, probably due to the presence of non-culturable strains. On the other hand, some species recovered as pure cultures from enrichments were not detected by direct DGGE analysis of total bacterial DNA from cassava dough, probably because such species were not dominant within the bacterial population.

The development of the LAB community in a commercial malt whisky fermentation was monitored by: (i) estimation of the total bacterial community viability by fluorescence microscopy; (ii) community fingerprinting by PCR-DGGE and DNA sequencing; (iii) indication of the relative activities of the major species present within the LAB community by RT-PCR-DGGE; and (iv) fermentation products accumulation. The evolution of the LAB community occurred in three broad phases. In an early stage yeasts prevailed, whereas LAB were present (especially S. thermophilus and heterofermentative lactobacilli) but inhibited; then the exponential growth of lactobacilli belonging to L. casei and L. fermentum occurred, which was accompanied by notable accumulation of lactic and acetic acids; finally, two new PCR bands were detected on DGGE gels, and the associated bacteria (L. acidophilus and L. delbrueckii) were largely responsible for the final phases of fermentation. This study suggested that community development in some systems can be thought of as an orderly sequence of chemical changes each catalysed by species waiting their turn in an imaginary eco-queue! Therefore, polyphasic ecology appears helpful in more predictable analysis of microbial population dynamics in complex, and often unpredictable, food fermentation processes.

## 5.13 ROLE OF MICRO-ORGANISMS IN THE FERMENTATION

### 5.13.1 Process

Initially, it might be hard to imagine how microorganisms contribute to the fermentation process. How can such small and microscopic cells unseen to the naked eyes produce such high volumes of products? How can such microorganisms produced such a rich diversity of fermentation products which are useful to human beings?

The answer to the first question is simple enough. One tiny microscopic microbial cell will not produce industrially significant volume of fermentation products. However, when we talk about billions and billions of microbial cells in the fermentation vat, the volume and concentration of fermentation products is industrially and economically significant!

The diversity of fermentation products produced by the microorganisms is attributed to the rich diversity of microorganisms which have a diverse metabolisms that can yield various types of fermentation products.

***Why Micro-Organisms are the Best Choice for Fermentation***

The success of using microorganisms for fermentation lies in their very microscopic and metabolic characteristics. It is good being small:

**1. High surface area to volume ratio:** Microorganisms are very very tiny creatures. Taking an example of a rod bacterium, we can see that it has six free surfaces that surrounds the bacterium. These six free surfaces interfaced with the surrounding environment from where they obtained their nutrients or to where they throw away their metabolic waste products.

With such a high number of free surface areas in a tiny volume of cell, it confers upon the bacterial cell a very high surface area to volume ratio. This very high surface area to volume ratio allows maximum or optimum surfaces for diffusions or molecular exchanges to occur between the microbial cell and the environment. No matter where the molecules are, they are easily accessible for diffusion into the microbial cell.

Compare this with the elephant which is such a huge animal. which has a very low ratio of surface area to volume ratio! If nutrients are to diffuse through the surface area of the elephant it would not be enough to supply every cell of the elephant.

Once the nutrient molecules diffuse through the cell wall and membrane, it can be easily transported to where its needed in the cell. The size of the cell is so microscopic that distance covered in the transportation of the molecules in the cell is a convenience.

With such an ease of diffusion of nutrient molecules from the environment into the cell, and the diffusion of waste products from within the cell outwards to the surrounding environment we can see that metabolism will be at optimum state.

The efficient nutrient uptake coupled with the small size of the cell will allow for rapid synthesis and reproduction of new cells. Microorganisms under ideal state will double up within hours. Animals like elephants and human may take months to reproduce themselves.

**2. Mode of nutrients transportation**: The nutrients which diffuse into the microbial can either use simple diffusion process which is powered by the differences in the concentration gradients between the environment and within the cell. For very small nutrient molecules, most would diffuse by the mechanism of passive diffusion. Larger and complex molecules use active or group transport which requires expansion or utilization of energy.

Microbes easily reproduce asexually! There is no real need to have a opposite partner cells, get married and reproduce. They will just as easily split their cells into two daughter cells which will later grow into larger cells and repeat the cycle.

**3. Genetic adaptability**: Microorganisms generally show the ability to adapt to new environment. They can get easily adapted to living under different environmental conditions and also adapting to new sources of carbon or substrate. This ability is the result of various genetic adaptation which selects "successful" strains through mutation and genetic recombination. Some of the bacteria are even equipped with plasmids which can synthesize new enzymes that help the microorganisms exploit the new environment. The very short generation times and the high population generation will aid the selection and recombination process.

**4. Metabolic diversity**: The unique thing about microbes are their metabolic diversity shown by various members of the microorganisms. They have the ability to use different energy sources and to use different types terminal electron acceptors.

Their ability to use different substrates is also correlated with the microbes ability to produce a diversity of fermentation products.

Fermentation bacteria are anaerobic, but use organic molecules as their final electron acceptor to produce fermentation end-products. Streptococcus, Lactobacillus, and Bacillus, for example, produce lactic acid, while Escherichia and Salmonella produce ethanol, lactic acid, succinic acid, acetic acid, $CO_2$, and $H_2$.

Fermenting bacteria have characteristic sugar fermentation patterns, i.e., they can metabolize some sugars but not others. For example, Neisseria meningitidis ferments glucose and maltose, but not sucrose and lactose, while Neisseria gonorrhoeaferments glucose, but not maltose, sucrose or lactose. Such fermentation patterns can be used to identify and classify bacteria.

During the 1860s, the French microbiologist Louis Pasteur studied fermenting bacteria. He demonstrated that fermenting bacteria could contaminate wine and beer during manufacturing, turning the alcohol produced by yeast into acetic acid (vinegar).

Pasteur also showed that heating the beer and wine to kill the bacteria preserved the flavor of these beverages. The process of heating, now called pasteurization in his honor, is still used to kill bacteria in some alcoholic beverages, as well as milk.

Pasteur described the spoilage by bacteria of alcohol during fermentation as being a "disease" of wine and beer. His work was thus vital to the later idea that human diseases could also be caused by microorganisms, and that heating can destroy them.

## 5.14 MICROBIAL STARTER CULTURES IN FERMENTATION

The success of many food and beverage based fermentation industries depends heavily on the availability of the microorganisms which drive the fermentation process. The types of microorganisms used, the quality and volume of inocula used in the fermentation process is crucial in the fermentation industries. In reality, most of the food or beverage based fermentation industries use pure or mix cultures for their fermentation depending on the type of fermentation carried out. Some fermentation requires the use of mono cultures while there are food or beverage fermentations that require the contribution of a number of species of microorganisms for the fermentation to be successful.

The microorganisms used in the various food and beverage fermentations were in fact historically isolated from the ' natural contaminants' which occur on the decomposing or naturally fermented foods.

The use of starter cultures is not new. In traditional fermentations traditional starter cultures have always been used in the fermentation though many do not realize it. Some of these starter cultures used are obtained from:

1. Remnants or residuals of past fermented food and beverage
2. Original flora that existed naturally on wrappings used in the fermentation such as banana leaves
3. Inocula that occur naturally in the substrate of the fermentation such as grapes used in wine making
4. Traditionally made 'kojis' used in many soy sauce or tuak fermentation

It is only now in the modern, sanitary, stainless steel fermentors of modern fermentation industries are ' starter cultures' used to start the desired fermentation process. Starter cultures are specially isolated and selected strains of microorganisms used in the fermentation which are massively cultivated under strict and scientific control to give the best and optimized fermentation which are not often seen in traditional food fermentations.

Before the first cultures entered the market, traditional fermented food producers tended to rely on their own cultures. However, inadequately characterised cultures and inadequately defined compositions generally variability in the quality of the fermented products which is regarded not satisfactory in the market.

The development of such cultures requires the expertise and experience of specialists microbiologists, biochemists and food technologists as not all combinations are suitable for all products. The use of different strains in the same microbial starters must not result negative influence on each other. Usually in any microbial starters it is not wise mix more than four different species in a culture,".

The advantages of using industrially prepared starter cultures for food and beverage fermentation are:

1. Use only the microorganisms responsible for the fermentation
2. No other microbial contaminants that can cause diversion from the optimal fermentation
3. Can use the best strains specifically designed for the particular fermentation
4. Use of large inocula of starter cultures will result in better and faster fermentation
5. Quality and stability of the fermentation will be maintained
6. Taste, texture, colour stability and shelf life are mainly determined by microorganisms involved in the maturation, i.e. fermentation, of the product

Fermentation technology is one of the oldest technologies employed in the food processing industry.

The microbial or animal cell obtains energy through glycolysis, splitting a sugar molecule and removing electrons from the molecules. Fermented products encompass, but are not limited to wine, beer, vinegar, bread, soy sauce, sauerkraut, kimchi, pickled, olive, different fermented milk products, a large number of cheeses and a variety of sausages.

Many species of microorganisms are used for carrying out the process of fermentation to produce useful products. They include bacteria, fungi, algae and actinomycetes.

Several advantage of fermentation technology included:

*Produces value added and add variety to the human being's diets.

*Preservation of the food.

*Food quality improvement through flavor development, nutrient enrichments.

*The food more nutritious.

*Detoxifies of foods.

Today large numbers of chemicals are produced by fermentation technology with the advent genetic engineering and the developments in computer technology.

### 5.14.1 Micro-Organisms and Fermentation Technology of Highland Barley Wine

Mold, yeast and lactic acid bacteria about Highland barley wine fermentation have been separated and identified on a large scale. By screening these microorganisms, two yeast strains ZY2 and ZY28, one mold strain ZM2 and one lactic acid bacteria strain ZB12 were separated as superior strains. Original strains ZM2 and ZY2 were breeded by protoptast induced mutation and DES mutagenesis. Pure cultured starters processing technology was achieved using optimum design of breeding technology by RSM. This write up provided a set of processing technique with the maintenance of general quality of traditional fermented barley wine which has subtle taste, low alcohol and typical flavour.

The main results are as following:

1. Quality and flavor characteristics of the barley wine and its starter were identified by analyzing Physiochemical Indexes, GC-MS and sensory index, comparing with Chinese Rice Wine and Japanese sake. Tibet barley wine starter was identified as Chinese yeast with different forms and qualities. The traditional barley wine was identified as fermentation wine and its physiochemical indexes of were between Rice Wine and Sake.
2. There are 48 flavor components in barley wine including 16 kinds of alcohol,5 kinds of ester, 6 kinds of acid, 10 kinds of aldehyde, 7 kinds of ketone and orther phenols.GC-MS shows that: highest content/threshold is aldehyde 41.76, then acetic acid 40.38, Isoamyl alcohol 13.38, caproic acid 4.65, ethyl lactate 3.07, Butyric acid 2.94, ethyl acetate 2.65. β-phenylethanol 2.14,3-hydroxy butanone1.36, these 9 compounds are most responsible for the flavor. Microorganisms of Barley wine starter and microorganisms flora during fermentation were studied using routine isolation, purification and identification method together with molecular biological technique, including PCR-Clone Analysis.

It shows that:

(1) The identification of dominant mold shows:there were 130 dominant mold strains, including 81 strains of Rhizopus.SPP:Rhizopus oryzae 39 strains, Rhizopus chinensis 15 strains, Rhizopus arrhizus 9 strains, 39 strains of Mucor SPP:including Mucor javanicus 18 strains, Mucor racermosus 11 strains, Mucor rouxians 8 strains, Absidia spp 8 strains. ZM2 was screened from all these mold strains with glucoamylase activity 520.27 mg/h•g. ZM2 is indentified as Rhizopus oryzae through ITS sequence analysis and Phylogenetic Tree Construction.

(2) The identification of dominant yeast shows:there are dominant yeast 580 strains, belonging to 6 species of Saccharomyces, Saccharomycopsis, Candida, Pichia, Debaryomyces, Cryptococcus, including 186 strains of Saccharomyces cerevisiae, Candida sake 38 strains, Pichia farinosa 16 strains, Debaryomyces hansenii 12 strains, Cryptococcus terreus 5 strains. ZY2 was screened from all these yeast strains for its high alcohol production ability, while ZY28 strain was screened by its ability to produce glucoamylase and good flavour. ZY2 and ZY28 are identified as Saccharomyces boulardii and Saccharomycopsis fibuligera, respectively through ITS sequence analysis and Phylogenetic Tree Construction.

(3) The identification of dominant lactic acid bacteria shows:38 strains in 3 species were dominant lactic acid bacteria, including Streptococcus faecalis, Pediococcus pentosaceus,Pediococcus acidialactici, Lactobacillus.johnsonii and Lactobacillus sake.ZB12 strain was screened for further research. It was identified as Pediococcus pentosaceus through ITS sequence analysis and Phylogenetic Tree Construction.3.ZY2 and ZM2 were breeded by protoplast induced mutation with UV and DES combination. .

It shows that:

(1) UV-Y2 was obtained with higher alcohol production by UV irradiation of ZY2 protoplast for 50s. The protoplast of UV-Y2 was treated with 1.5%DES for 30 minutes, then the final mutant strain UV-DES-Y2 was obtained which can produce more than 33.33% alcohol after a series of screening by TTC plating screen, Durham fermentation and erlenmeyer flask fermentation.etc.

(2) UV-M4 was obtained with higher alcohol production by UV irradiation of ZM2 protoplast for 30s. The protoplast of UV-Y2 was treated with 0.9%DES 50 minutes, then the final mutant strain UV-DES-M4 was obtained which can produce more than 50.9% Glucoamylase. The optimum of enzymatic production for starter 1 is achieved by Using RSM:highland barley powder with 40% water, 8.06% soya flour, 1.0% inoculation, cultivating at $30.05^0$for 42 h and the Glucoamylase is 895.32 mg/h•g dry starter.

(3) The optimal processes of making yeast solid starter:highland barley powder 50% glucose-bean sprout liquid medium (GBM) whose initial pH is 3.5( for ZY2) or 5 (for ZY28), 15% inoculating ratio which were previously breeded in GBM for 24h, cultivating at $32^0$c (for ZY2) or $25^0$C (for ZY28) for 30h, so the starter 2 (ZY2) and starter 3(ZY28) were obtained.

(4) The optimization experiments of brewing technique with pure strain fermentation were carried with the following results. Using Tibet barley 320 as raw materials,soaked or no soaked with the ratio of 1:2 water,steamed 2h,with the inoculum's concentration of 2% pure starter 1,2% pure starter 2 and1% pure starter 3, fermented on the condition of $28^o$Cand 72h.

(5) The change of compositions of the barley wine with pure strain and the traditional barley wine during its fermentation process are studied. The results were showed as follows: The changing tendency of compositions during two kinds of fermentation process was similar:pH declined, temperature increase slowly and declined slowly with the highest temperature 36º. The activities of the saccharifying enzymes and the reducing sugar content increased significantly (P<0.05) during the first period of fermentation to 24h, then decreased gradually. The total sugar content decreased significantly (P<0.05) during all the fermentation while the alcoholicity increased significantly (P<0.05). The alcoholicity during the fermentation with

pure strain process reached to 10.3%higher than traditional fermentation. Both the ammonia-N and the acid prote in ase increased significantly ($P<0.05$) up to 24 h of fermentation; and then decreased slowly. The total content of barley wine during the fermentation with pure strain reached to 80.923 mg/100 ml without Methionine.

## 5.15 FERMENTATION TECHNOLOGY: COOKING UP NEW THERAPEUTICS

Therapeutic drugs are substances used for health restoration in the treatment of disease, drawing on the utilization of microbial bio-factories. These bio-factories offer the pharmaceutical industry a reliable, powerful and economic source for therapeutics. In order to demonstrate the diversity of applications for bio-factories, Silvian Shama, Business Development Manager at the Jerusalem plant, looks at examples of therapeutic drugs that are primarily isolated through microbial fermentation.

### 5.15.1 Chemical Therapeutics

The term chemical therapeutics refers to toxic small organic molecules with a molecular weight of 1000-2000 which are produced naturally by microorganisms in response to stress and are called secondary metabolites. These molecules may be complex in nature, such as heterocyclics, with several chiral centers, large stereospecific rings or a unique conjugated double bond system. Small, cytotoxic molecules are employed in anti-cancer therapy (chemotherapy) and also in the treatment of infectious diseases (antibiotics).

**Cancer therapy:** The anti-cancer therapeutic activity of small molecule drugs in clinical use is limited by their general toxicity to proliferating cells, including some normal cells. In the lack of specificity and effective targeting, chemotherapy using High Potency Active Pharmaceutical Ingredients (HPAPIs) involves high doses and often severe side effects. Over the last few years the pharmaceutical industry has taken a new approach to targeted drug delivery by conjugating HPAPIs to monoclonal antibodies creating highly specific 'smart bombs' which can be directed to cancer cells through highly expressed antigens. This means that not only are smaller quantities of a cytotoxic drug required, but non-target cells are spared many of the cytotoxic effects. Various HPAPI Antibody Drug Conjugates (ADCs) are currently in clinical trials with major pharmaceutical companies and some are gaining FDA approval for difficult-to-treat indications, mainly in the treatment of cancer. Example – Wyeth's FDA approved Mylotarg®

ADC consists of a recombinant humanized IgG4 kappa antibody, conjugated with a cytotoxic anti-tumor antibiotic, calicheamicin, which is isolated through the fermentation of a bacterium, Micromonospora echinospora subspecies calichensis. The antibody portion of Mylotarg binds specifically to the CD33 antigen, a sialic acid-dependent adhesion protein which is found on the surface of leukemic blasts and immature normal cells of myelomonocytic lineage, but not on normal hematopoietic stem cells.

**Treatment of infectious diseases:** Antibiotics are substances or compounds that kill or inhibit the growth of microorganisms, mainly bacteria. Many of the antibiotics are sourced from living microorganisms, used either in their natural state or in the case of semi-synthesis, chemically modified. Semi–synthesis is usually used when the precursor molecule is too structurally complex, too costly or too inefficient to be produced by total synthesis. It is also possible that the semi synthetic derivative exhibits better performance than the original bio molecule itself in terms of potency, stability or safety.

Example: Aminoglycosides are produced and isolated from bacteria of the Streptomyces genus, used in their existing form to provide drugs such as Streptomycin, a bactericidal agent and Hygromycin B, a killer of bacteria and fungi.

### 5.15.2 Biological Therapeutics

The term biological therapeutics refers to large organic molecules, mainly proteins. Proteins are polymers of amino acids arranged in a linear chain and folded into a globular form. The sequence of an amino acids protein is defined by the sequence of a gene, which is encoded in the genetic code. Biological therapeutics are isolated from the native producing microorganisms or, more often, expressed in a heterologous microorganism using recombinant DNA technology (recombinant proteins). Fermentation derived biological therapeutics have been employed in a wide range of medical indications.

Native protein example – Collagenase - Collagenases are a group of enzymes produced by the bacterial pathogen Clostridium genus. Upon infection, Collagenases break the collagen peptide bond, a key component of the animal extracellular matrix. Destruction of the extracellular structures causes severe tissue damage and can facilitate the spread of serious medical conditions such as gaseous gangrene.

Essentially turning an enemy into a friend, native isolated collagenases are employed in a variety of medical indications involving collagen disorders.

Collagenases approved for medical uses are: Santyl® ointment and Iruxol, for the removal of dead skin from wounds and burned areas in poorly healing wounds and necroses, and Xiafles Dupuytren's contracture, a condition resulting in the contracture of the fingers into the palm. Other collagenases are in currently in clinical trials for indications such as:

- **Frozen shoulder**: An inflammation and thickening of the shoulder capsule;
- **Cellulite: A** condition characterized by dimpling of the skin typically affecting the thighs and buttocks; and
- **Lipomas:** Benign fatty tumors that occur as bulges under the skin.

Some approved products are already on the market such as Genentech's Lucentis® (ranibizumab injection), which is a recombinant humanized IgG1 kappa iso type monoclonal antibody fragment designed for intraocular use.

Applications of microbial fermentation technology in new therapeutics are expanding and SAFC offers unique cGMP fermentation-based manufacturing capabilities through its Jerusalem facility, including high potent APIs and biologics. SAFC scientific teams provide a complete range of services, including process evaluation and development, optimization and scale up, technology transfer, manufacturing, analytical testing and regulatory filing to help customers bring new drugs to market faster.

## 5.16 THE DEVELOPMENT OF DEEP-TANK FERMENTATION

During World War II the governments of the United States and the United Kingdom approached the largest U.S. chemical and pharmaceutical companies to enlist them in the race to mass produce penicillin, the "wonder drug." One of these companies, Pfizer, already had experience with fermentation techniques, first implemented twenty years earlier to manufacture citric acid. Building on that experience, Pfizer succeeded in producing large quantities of penicillin using deep-tank fermentation. Its success helped make penicillin available to Allied soldiers by the end of the war.

The American Chemical Society designated the development of deep-tank fermentation by Pfizer Inc as a National Historic Chemical Landmark on June 12, 2008 in Brooklyn, New York.

### 5.16.1 The Penicillin Problem

The story of the accidental discovery of penicillin by Alexander Fleming has been told often. Fleming was already well known for his earlier work on staphylococci when, upon returning from a long holiday in September 1928, he noted that one petri dish contained colonies of Staphylocccus except for clear area contaminated with a fungus that appeared to inhibit bacterial growth. Fleming identified the contaminant as a strain of Penicillium and he found that it killed a host of Gram-positive bacteria, including those that caused scarlet fever, pneumonia, gonorrhea, meningitis, and diphtheria.

Cultivating penicillin proved difficult and the small quantities that could be produced were unstable. Fleming also believed that penicillin would not remain in the human body long enough to kill bacteria. Taken together, these facts convinced Fleming that penicillin would never be an important antibiotic for treating infections. For the next decade, penicillin remained a laboratory curiosity, until Howard Florey and a team of researchers at Oxford University demonstrated its potential life-saving properties. The researchers also discovered a method to purify penicillin and to keep it in a stable form. By 1940, clinical trials were underway and their success led to the hunt for a method to manufacture this "wonder drug." But by then England was at war, making mass production difficult.

### 5.16.2 Pfizer and Fermentation

The answer begins with a small chemical company that would grow into a pharmaceutical giant.

Charles Pfizer and Charles Erhart, cousins and recent émigrés from Germany, established Charles Pfizer & Company in the Williamsburg neighborhood of Brooklyn in 1849. The company's first product cured a common 19th-century malady, intestinal worms, which were usually treated with santonin, an anti-parasitic so bitter that most people thought the cure worse than the condition. But the cousins, pooling the skills of Pfizer, a chemist, and Erhart, a confectioner, blended santonin with an almond-toffee flavoring to create a candy-flavored medicine that patients would take.

Pfizer's initial success with santonin encouraged the cousins to look for other opportunities to manufacture "fine chemicals." Among the company's products were quinine, used as an analgesic; borax, used as a preservative and a laundry detergent; boric acid, a topical antiseptic; cream of tartar, used in baking powder, and tartaric acid, also used in foods (both by-products of the European wine industry); Rochelle salts, a laxative and diuretic; camphor, a painkiller; and strychnine, a tonic and appetite stimulant. During the Civil War Pfizer, like other chemical and pharmaceutical companies, responded to the demand for painkillers, preservatives, and disinfectants, producing many products with medicinal applications, including iodine, morphine, and chloroform.

Pfizer's initial blockbuster product was citric acid, which it added to its product line in the 1880s. Citric acid is used in foods and beverages – notably soft drinks – because it is a natural preservative which can also add an acidic, slightly sour, taste. By supplying citric acid for new, popular drinks like Coca-Cola, Pepsi-Cola, and Dr. Pepper, Pfizer was able to dominant the market. At first, the company used the standard production methods, obtaining citric acid from unripe fruit from Italy, California, Florida, and the West Indies, and treating it with lime (CaO) and sulfuric acid to yield a solution of crystallized citric acid.

The botanist Carl Wehmer discovered in 1893 that the Penicillium mold could produce citric acid from sugar. Later, J.A. Martin discovered that fermenting sugar could yield citric acid. But these were ideas ahead of their time because no one knew how to manufacture citric acid from these sources on a commercial scale. That is, until James Currie, a food chemist, discovered that citric acid could be fermented from certain strains of the mold Aspergillis niger combined with sugar.

Currie later described how he took his discovery to Pfizer in 1917: "I met with John Anderson, who was Pfizer's chairman of the board. During our first meeting, Mr. Anderson introduced me to another Pfizer official with the remark, 'Dr. Currie is up here now and I think he has something interesting." Pfizer officials quickly realized that the "something interesting" was the prospect of producing vast quantities of citric acid from sugar, rather than from imported citrus fruit.

The fact that Currie joined Pfizer during World War I was no coincidence. Pfizer officials, led by John Anderson, realized that the company had to reduce its vulnerability as an importer of raw materials. The war cut off the supply of Italian citrus, and all raw material deliveries from Europe were threatened by the German submarine blockade. Supplying medicines and chemicals for the war helped the company survive, but overall sales declined. Now, more than ever, Pfizer needed to find ways to produce citric acid without using citrus.

Currie plugged on, turning to Gordon Dryer machines, which controlled humidity and temperature. He also experiment with different-sized pans and their depth and with temperatures. In 1919 Pfizer opened a pilot plant using his fermentation process, named SUCIAC – "Sugar Under Conversion to Citric Acid." Still, yields were disappointingly low for a time, so Currie kept fine tuning the process – using smaller pans and a better ventilation system – to increase output. By 1924 growing yields of citrus acid convinced Pfizer officials to build a SUCIAC plant. The new structure began operating in 1926; that year the output of citric acid using fermentation technology far outpaced the production based on lemons and limes. By the end of the decade Pfizer was dominating the market and in 1929 the company reached a milestone; it produced 5.9 million pounds of citric acid without using any lemon juice, lime juice, or foreign citrate of lime in its production process.

### 5.16.3 Deep-Tank Fermentation

Pfizer's experience with fermentation in the production of citric acid provided valuable lessons for the company when biochemistry created new germ-killing antibiotics, many of which could be produced in the laboratory but not on a commercial scale. And Pfizer gained even more experience as the company searched for ways to use fermentation to produce other chemicals or to improve existing methods. For example, in 1933, Jasper Kane substituted molasses, a cheap by-product of sugar refining, for sugar in the fermentation of citric acid.

Pfizer soon was producing other products by fermentation, most notably gluconic acid ($C_6H_{12}O_7$), used as an additive in food to regulate acidity and as a cleaning agent. Unlike the case of citric acid, an important product that Pfizer sought to produce more efficiently, Pfizer succeeded in marketing gluconic acid only after developing a commercial process involving deep-tank fermentation. Pfizer had tried unsuccessfully to use deep tanks to produce citric acid; in 1929 the company succeeded in producing gluconic acid in a submerged aerobic medium in stirred deep tanks which controlled for pH and the sterility of the air.

In following years the company used fermentation processes to produce itaconic acid, fumaric acid, and intermediates for chemical synthesis. The production of gluconic acid and other products using deep-tank fermentation was a significant breakthrough for Pfizer. While the company used fermentation techniques in manufacturing citric acid, it was the development of deep tanks for gluconic acid that paved the way for the successful mass production of penicillin.

In the 1930s Pfizer, which had close ties with the pharmaceutical and food industries, became interested in the commercial production of vitamins, beginning with vitamin C. Ascorbic acid was first isolated by Albert Szent-Györgyi in 1928; shortly after, its structural formula was established and a method of synthesis discovered. 1936 Pfizer began marketing synthetic vitamin C made with submerged fermentation as the first step in the process. Pfizer soon became the leading manufacturer, and the company decided to market other vitamins, introducing vitamin B2 in 1938 and, after the war, vitamin B12.

### 5.16.4 Penicillin Solved

In 1941, the British government sought help from the U.S. scientific community in solving the problem of mass producing penicillin. Four chemical and pharmaceutical companies, Pfizer included, responded, plunging into the race to produce the world's first "wonder drug."

In early 1940s Jasper Kane and his colleagues increased penicillin production incrementally as well as the drug's potency and purity, but the gains were frustratingly slow. At first, Kane used flasks and pans similar to the fermentation techniques for citric acid. In 1942, Kane suggested switching to the deep tanks that worked so well for gluconic acid. Kane's suggestion was risky for it meant Pfizer would have to curtail production of other, more profitable products while it concentrated on penicillin. As one executive, John Smith, said: "The mold is as temperamental as an opera singer, the yields are low, the isolation murder, the purification invites disaster. Think of the risks!"

The company decided to take the gamble. In September 1943 it purchased the old Rubel Ice Plant, a nearby building that had the refrigeration equipment required, rebuilding it into the world's first large-scale penicillin factory. On March 1, 1944 Pfizer's penicillin plant opened. It contained fourteen 7,500-gallon tanks; soon the company was producing five times more penicillin then originally estimated, making Pfizer the leading supplier of the drug; most of the penicillin that went ashore with Allied forces on D-Day came from this plant on Marcy Avenue in Brooklyn.

Behind the numbers stood a marvel of chemical engineering. Production began with a sterile culture of the penicillin mold, which then was propagated, first in three-liter flasks, next in 200-gallon "seed" tanks. The culture then moved to huge fermenter tanks containing microbe fodder, chiefly corn steep liquor, milk sugar, salts, and minerals. The mold was allowed to grow for two to four days.

The trickiest part (the whole process was tricky) was extracting penicillin from the broth, for penicillin was a very stingy "magic bullet:" only four parts drug per 10,000 parts broth. The extracted material was then purified and bottled in sterile rooms as extreme care had to be taken in these last steps to avoid contamination. After the bottles were filled, the penicillin moved through a freezing apparatus and then into a vacuum drier which dehydrated the drug.

Pfizer used the lessons learned to develop other antibiotics, first applying its fermentation techniques to the manufacture of streptomycin, a drug developed by Selman Waksman and his staff at Rutgers. In 1949 Pfizer scientists, after testing tens of thousands of soil samples, found a micro-organism in soil from America's Midwest that proved effective against a wide range of deadly bacteria. Terramycin, derived from the Latin for "earth fungus," was the first antibiotic developed exclusively by Pfizer scientists.

Alexander Fleming and the discoverers of other antibiotics deserve all the credit they have won. But their work would have remained on the laboratory shelf if not for the development of deep-tank fermentation. As David Wilson wrote in his book In Search of Penicillin (Knopf, 1976), "It is the biggest single failing of the myth about penicillin that it ignores the technological breakthrough of deep fermentation, a breakthrough that was every bit as vital to the successful development of penicillin as any of the more dramatic laboratory work."

In the early 20th century Pfizer developed innovative fermentation technology, applying it first to the mass production of citric acid. In subsequent years, under the direction of James Currie and Jasper Kane, Pfizer perfected deep-tank fermentation, an aseptic process for growing large quantities of microorganisms which require oxygen for survival. When scientists in England were unable to produce penicillin on a large scale during World War II, Kane suggested trying deep-tank fermentation. In a major feat of chemical engineering, the company rebuilt an old ice plant, which had the refrigeration machinery required for submerged fermentation, and opened the world's first large-scale penicillin facility on March 1, 1944. Pfizer manufactured other antibiotics, notably Terramycin, and vitamins using deep-tank fermentation techniques.

## 5.17 ROLE OF YEAST IN FERMENTATION PROCESS

Yeasts are eukaryotic micro-organisms classified in thekingdom Fungi, with 1,500 species currently described estimated to be only 1% of all fungal species. Most reproduce asexually by mitosis, and many do so via an asymmetric division process called budding. Yeasts are unicellular, although some species with yeast forms may become multicellular through the formation of a string of connected budding cells known as pseudohyphae, or false hyphae, as seen in most molds. Yeast size can vary greatly depending on the species, typically measuring 3–4 µm indiameter, although some yeasts can reach over 40 µm.

By fermentation the yeast species Saccharomyces cerevisiaeconverts carbohydrates to carbon dioxide and alcohols - for thousands of years the carbon dioxide has been used in bakingand the alcohol in alcoholic beverages. It is also extremely important as a model organism in modern cell biology research, and is one of the most thoroughly researched eukaryoticmicroorganisms. Researchers have used it to gather information about the biology of the eukaryotic cell and ultimately human biology. Other species of yeast, such as Candida albicans, are opportunistic pathogens and can causeinfections in humans. Yeasts have recently been used to generate electricity in microbial fuel cells, and produce ethanol for the biofuel industry.

### 5.17.1 Nutrition and Growth

Yeasts are chemoorganotrophs, as they use organic compounds as a source of energy and do not require sunlight to grow. Carbon is obtained mostly from hexose sugars, such as glucose and fructose, or disaccharides such as sucrose and maltose. Some species can metabolize pentosesugars like ribose, alcohols, and organic acids. Yeast species either require oxygen for aerobiccellular respiration (obligate aerobes), or are anaerobic, but also have aerobic methods of energy production (facultative anaerobes). Unlike bacteria, there are no known yeast species that grow only anaerobically (obligate anaerobes). Yeasts grow best in a neutral or slightly acidic pH environment.

Yeasts vary in what temperature range they grow best. For example, Leucosporidium frigidum grows at -2 to 20 °C (28 to 68 °F), Saccharomyces telluris at 5 to 35 °C (41 to 95 °F) and Candida slooffi at 28 to 45 °C (82 to 113 °F). The cells can survive freezing under certain conditions, with viability decreasing over time.

Yeasts are generally grown in the laboratory on solid growth media or in liquid broths. Common media used for the cultivation of yeasts include potato dextrose agar (PDA) or potato dextrose broth, Wallerstein Laboratories nutrient (WLN) agar, yeast peptone dextrose agar (YPD), and yeast mould agar or broth (YM). Home brewers who cultivate yeast frequently use dried malt extract (DME) andagar as a solid growth medium. The antibiotic cycloheximide is sometimes added to yeast growth media to inhibit the growth of Saccharomyces yeasts and select for wild/indigenous yeast species. This will change the yeast process.

The appearance of a white, thready yeast, commonly known as kahm yeast, is often a byproduct of the lactofermentation (or pickling) of certain vegetables, usually the result of exposure to air. Although harmless, it can give pickled vegetables a bad flavour and so must be removed regularly during fermentation.

### 5.17.2 History

The word "yeast" comes to us from Old English gist, gyst, and from the Indo-European root yes-, meaning boil, foam, or bubble. Yeast microbes are probably one of the earliest domesticated organisms. People have used yeast for fermentation and baking throughout history. Archaeologists digging in Egyptian ruins found early grinding stones and baking chambers for yeasted bread, as well as drawings of 4,000-year-old bakeries and breweries. In 1680, the Dutch naturalist Anton van Leeuwenhoek first microscopically observed yeast, but at the time did not consider them to be living organisms, but rather globular structures. In 1857, French microbiologist Louis Pasteur proved in the paper "Mémoire sur la fermentation alcoolique" that alcoholic fermentation was conducted by living yeasts and not by a chemical catalyst. Pasteur showed that by bubbling oxygen into the yeast broth, cell growth could be increased, but fermentation was inhibited - an observation later called the "Pasteur effect".

By the late 18th century, two yeast strains used in brewing had been identified: Saccharomyces cerevisiae, so called top fermenting yeast, and S. carlsbergensis, bottom fermenting yeast. S. cerevisiae has been sold commercially by the Dutch for bread making since 1780; while around 1800, the Germans started producing S. cerevisiae in the form of cream. In 1825 a method was developed to remove the liquid so the yeast could be prepared as solid blocks. The industrial production of yeast blocks was enhanced by the introduction of the filter press in 1867. In 1872, Baron Max de Springer developed a manufacturing process to create granulated yeast, a technique that was used until the first World War. In the United States, naturally occurring airborne yeasts were used almost exclusively until commercial yeast was marketed at the Centennial Exposition in 1876 in Philadelphia, where Charles L. Fleischmann exhibited the product and a process to use it, as well as serving the resultant baked bread.

### 5.17.3 Reproduction

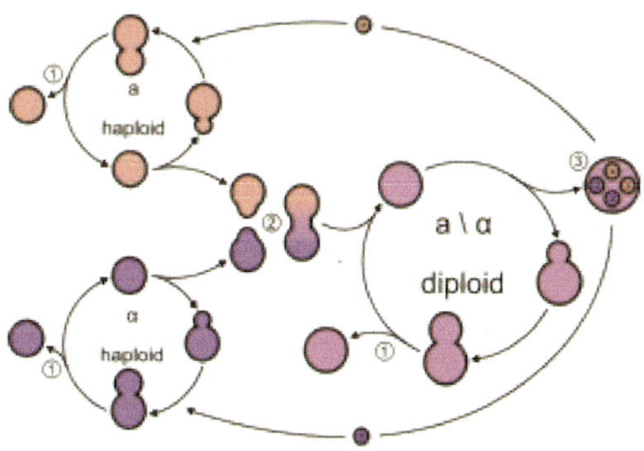


**Fig. 2. Reproduction**

**The yeast cell's life cycle:**

1. Budding
2. Conjugation
3. Spore

Yeasts, like all fungi, may have asexual and sexualreproductive cycles. The most common mode of vegetative growth in yeast is asexual reproduction by budding. Here a small bud (also known as a bleb), or daughter cell, is formed on the parent cell. The nucleus of the parent cell splits into a daughter nucleus and migrates into the daughter cell. The bud continues to grow until it separates from the parent cell, forming a new cell. Some yeasts, including Schizosaccharomyces pombe, reproduce by fission instead of budding.

Under high stress conditions, haploid cells will generally die; under the same conditions, however, diploid cells can undergo sporulation, entering sexual reproduction (meiosis) and producing a variety of haploid spores, which can go on to mate (conjugate), reforming the diploid.

Some pucciniomycete yeasts, particularly species of Sporidiobolus and Sporobolomyces produce aerially dispersed, asexual ballistoconidia.

### 5.17.4 Uses

The useful physiological properties of yeast have led to their use in the field of biotechnology. Fermentation of sugars by yeast is the oldest and largest application of this technology. Many types of yeasts are used for making many foods: baker's yeast in bread production; brewer's yeast in beerfermentation; yeast in wine fermentation and for xylitol production. So-called red rice yeast is actually a mold, Monascus purpureus. Yeasts include some of the most widely used model organisms for genetics and cell biology.

## 5.18 ALCOHOLIC BEVERAGES

Alcoholic beverages are defined as beverages that contain ethanol ($C_2H_5OH$). This ethanol is almost always produced by fermentation – the metabolism of carbohydrates by certain species of yeast under anaerobic or low-oxygen conditions. Beverages such as wine, beer, or distilled spirits all use yeast at some stage of their production. A distilled beverage is a beverage containing ethanol that has been purified by distillation. Carbohydrate-containing plant material is fermented by yeast, producing a dilute solution of ethanol in the process. Spirits such as whiskey and rum are prepared by distilling these dilute solutions of ethanol. Components other than ethanol are collected in the condensate, including water, esters, and other alcohols, which account for the flavour of the beverage.

### 5.18.1 Beer

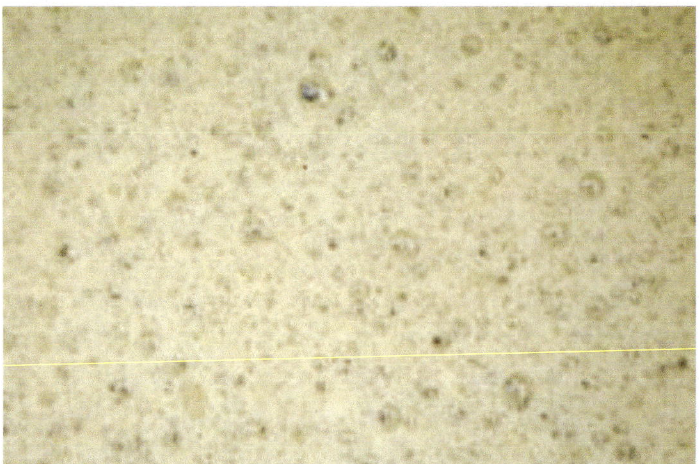

**Fig. 3. Beer Being Fermented by Brewers Yeast.**

Brewing yeasts may be classed as "top cropping" (or "top fermenting") and "bottom cropping" (or "bottom-fermenting"). Top cropping yeasts are so called because they form a foam at the top of the wort during fermentation. An example of a top cropping yeast is Saccharomyces cerevisiae, sometimes called an "ale yeast". Bottom cropping yeasts are typically used to produce lager-type beers, though they can also produce ale-type beers. These yeasts ferment well at low temperatures. An example of bottom cropping yeast is Saccharomyces pastorianus, formerly known as S. carlsbergensis.

Decades ago, taxonomists reclassified S. carlsbergensis as a member of S. cerevisae, noting that the only distinct difference between the two is metabolic. Lager strains of S. cerevisae secrete an enzyme called melibiase, allowing it to hydrolyse the disaccharide melibiose into more fermentable monosaccharides. Top cropping and bottom cropping, cold fermenting and warm fermenting distinctions are largely generalizations used to by the laymen to communicate to the general public. For more on the taxonomical differences, see Terrance M. Dowhanick, B.Sc., Ph. D "Yeast - Strains and Handling Techiques" in The Practical Brewer, a publication of the Master Brewers Association of the Americas.

The most common top cropping brewer's yeast, S. cerevisiae, is the same species as the common baking yeast. Brewer's yeast is also very rich in essential minerals and the B vitamins (except B12). However, baking and brewing yeasts typically belong to different strains, cultivated to favour different characteristics: baking yeast strains are more aggressive, to carbonate dough in the shortest amount of time possible; brewing yeast strains act slower, but tend to produce fewer off-flavours and tolerate higher alcohol concentrations (with some strains, up to 22%).

Dekkera/Brettanomyces is a genus of yeast known for their important role in the production of Lambicand specialty sour ales, along with the secondary conditioning of a particular Belgian Trappist beer. The taxonomy of the genus Brettanomyces has been debated since its early discovery and has seen many re-classifications over the years. Early classification was based on a few species that reproduced asexually through multipolar budding. Shortly after, the formation of ascospores was observed and the genus Dekkera, which reproduces sexually, was introduced as part of the taxonomy. The current taxonomy includes five species within the genera of Dekkera/Brettanomyces. Those are the anamorphs Brettanomyces bruxellensis, Brettanomyces anomalus, Brettanomyces custersianus, Brettanomyces naardenensis, and Brettanomyces nanus, with teleomorphs existing for the first two species, Dekkera bruxellensis and Dekkera anomala. The distinction between Dekkera and Brettanomyces is arguable with Oelofse et al. Citing Loureiro and Malfeito-Ferreira from 2006 when they affirmed that current molecular DNA detection techniques have uncovered no variance between the anamorph and teleomorph states. Over the past decade, Brettanomyces spp. have seen an increasing use in the craft-brewing sector of the industry with a handful of breweries having produced beers that were primary fermented with pure cultures of Brettanomyces spp. This has occurred out of experimentation as very little information exists regarding pure culture fermentative capabilities and the aromatic compounds produced by various strains. Dekkera/Brettanomyces spp. have been the subjects of numerous studies conducted over the past century although a majority of the recent research has focused on enhancing the knowledge of the wine industry. Recent research on 8 Brettanomyces strains available in the brewing industry focused on strain specific fermentations and identified the major compounds produced during pure culture anaerobic fermentation in wort.

### 5.18.2 Wine

**Fig. 4. Fresh grapes with visible bloom**

Yeast is used in winemaking, where it converts the sugars present in grape juice (must) into ethanol. Yeast is normally already present on grape skins (the white powder called "the bloom"). Fermentation can be done with this endogenous "wild yeast," but this procedure gives unpredictable results, which depend upon the exact types of yeast species present. For this reason, a pure yeast culture is usually added to the must; this yeast quickly dominates the fermentation. The wild yeasts are repressed, which ensures a reliable and predictable fermentation.

Most added wine yeasts are strains of S. cerevisiae, though not all strains of the species are suitable. Different S. cerevisiae yeast strains have differing physiological and fermentative properties, therefore the actual strain of yeast selected can have a direct impact on the finished wine. Significant research has been undertaken into the development of novel wine yeast strains that produce atypical flavour profiles or increased complexity in wines.

The growth of some yeasts, such as Zygosaccharomyces and Brettanomyces, in wine can result inwine faults and subsequent spoilage. Brettanomyces produces an array of metabolites when growing in wine, some of which are volatile phenolic compounds. Together, these compounds are often referred to as "Brettanomyces character", and are often described as "antiseptic" or "barnyard" type aromas. Brettanomyces is a significant contributor to wine faults within the wine industry.

## 5.18.3 Baking

Yeast, most commonly S. cerevisiae, is used in baking as a leavening agent, where it converts thefermentable sugars present in dough into the gas carbon dioxide. This causes the dough to expand or rise as gas forms pockets or bubbles. When the dough is baked, the yeast dies and the air pockets "set", giving the baked product a soft and spongy texture. The use of potatoes, water from potato boiling, eggs, or sugar in a bread dough accelerates the growth of yeasts. Most yeasts used in baking are of the same species common in alcoholic fermentation. Additionally, Saccharomyces exiguus, a wild yeast found on plants, fruits, and grains, is occasionally used for baking. Sugar and vinegar provide the best conditions for yeast to ferment. In bread making, the yeast initially respires aerobically, producing carbon dioxide and water. When the oxygen is depleted,anaerobic respiration begins, producing ethanol as a waste product; however, this evaporates during baking.

**Fig. 5. A block of fresh yeast**

It is not known when yeast was first used to bake bread. The first records that show this use came from Ancient Egypt. Researchers speculate a mixture of flour meal and water was left longer than usual on a warm day and the yeasts that occur in natural contaminants of the flour caused it to ferment before baking. The resulting bread would have been lighter and tastier than the normal flat, hard cake.

**Fig. 6. Active dried yeast, a granulated form in which yeast is commercially sold**

Today, there are several retailers of baker's yeast; one of the best-known in North America is Fleischmann's Yeast, which was developed in 1868. During World War II, Fleischmann's developed a granulated active dry yeast, which did not require refrigeration and had a longer shelf lifethan fresh yeast. The company created yeast that would rise twice as fast, reducing baking time. Baker's yeast is also sold as a fresh yeast compressed into a square "cake".

This form perishes quickly, and must therefore be used soon after production. A weak solution of water and sugar can be used to determine if yeast is expired. In the solution, active yeast will foam and bubble as it ferments the sugar into ethanol and carbon dioxide. Some recipes refer to this as proofing the yeast as it "proves" the viability of the yeast before the other ingredients are added. When using a sourdough starter, flour and water are added instead of sugar; this is referred to as proofing the sponge.

When yeast is used for making bread, it is mixed with flour, salt, and warm water or milk. The dough is kneaded until it is smooth, and then left to rise, sometimes until it has doubled in size. Some bread doughs are knocked back after one rising and left to rise again. A longer rising time gives a better flavour, but the yeast can fail to raise the bread in the final stages if it is left for too long initially. The dough is then shaped into loaves, left to rise until it is the correct size, and then baked. Dried yeast is usually specified for use in a bread machine; however, a (wet) sourdough starter can also work.

## 5.19 YEAST FERMENTATION TECHNOLOGIES

The yeast strain, Saccharomyces cerevisiae, is a species of budding yeast commonly used as a leavening in baking and a microorganism for fermentation. The strain of S. cerevisiae is known as a model organism for the study of human biology providing valuable information related to cell growth and genetics. Additionally, the microorganism supports industrial applications as the leading fermenting yeast in the biological and medicinal sciences producing commercially available proteins-enzymes and co-enzymes.

### 5.19.1 History of Yeast Fermentation Technology

Oriental Yeast Company (OYC) was established in 1929 as the first company in Japan to produce yeast for baking. As the need for yeast grew both in Japan and worldwide, OYC grew as well-producing yeast for several applications including baking powders, laboratory animal feeds, and aquacultural feeds. OYC also focused on the establishment of yeast-derived products for medical diagnostic applications including enzymes, coenzymes, antibodies, recombinant proteins, and clinical chemistry reagents.

**Fig. 7. Oriental yeast company.**

Committed to quality products and customer satisfaction, OYC has continued to grow to better serve its global market. Strictly maintaining its ISO 9000 accreditation, it has entered into several partnerships and technology ventures worldwide-establishing offices in the United States in 1994 as well as Europe and China in 2008. In addition to yeast for baking and bakery ingredients, OYC has served diagnostic, pharmaceutical, and nutritional industries as well as contract labs, research institutes and universities for over 30 years.

OYC's worldwide network of products-supported by close client relationships and excellent customer service-ensures on time shipments, up-to-date technical information, and complete sales support.

Setting benchmarks for best practices and industry standards has been an important part of OYC's history. OYC's two research ventures focus on both the nutritional and scientific applications of yeast fermentation science:

**Fig. 8. Laboratory of yeast and fermentation.**

Tokyo Laboratory of Yeast and Fermentation was founded in tandem with OYC in 1929. Ever since, the laboratory has played a large role in the development and research of baker's yeast in Japan. The Tokyo Laboratory has isolated over 13,000 strains of bakers' yeast and helped bring 30 varieties to the world market.

Nagahama Institute for Biochemical Science was founded in 1987 to establish research and development, putting gene technology and other related technologies to practical use for analysis, testing, and production. With diverse areas of research-antiserum studies based on recombinant gene technology, protein engineering, and gene analysis-the Institute developed a series of raw materials and procedures that are used worldwide.

### 5.19.2 Utilisation of Yeast for Baking

- Bread came into being from observations made by the Egyptians 3,000 years ago of the transformation of the dough by beer froth which they scooped up from the surface of bowls. Since then, the history of bread, yeast and men is tightly linked. For a long time, bread was a basic food element in Europe and in the whole Mediterranean basin, with a high symbolic content.
- Without fermentation by yeast, a living and natural product, we could not obtain the equivalent of bread or Viennese pastry.
- Bread is less important nowadays as a basic food in developed countries, but it is still a healthy product, harmoniously helping to bring about a balanced diet and nutritionists recommend its consumption.
- Baking fermentation
- In the process of fermentation, yeast produces carbon dioxide, alcohol and other compounds which enable dough to rise and modify its physical properties. When the fermentation is correctly achieved, depending on the quality of flour among other things, the baker will obtain the proper external and internal characteristics (grain and texture) suitable for a determined finished product.
- Together with alcohol, a small amount of other volatile compounds are released. These are involved in the original taste and flavours of baked leavened products. Most of these compounds, and all of the alcohol, evaporate when baked.
- There are two stages during the fermentation of a dough made from flour, water, salt and yeast, and with no extra sugar added.
- First of all, yeast ferments sugars naturally present in flour, which can be directly and easily assimilated. These sugars represent about 1.5% of the flour weight. At the end of this first stage, gaseous releases more or less slow down.

- The second stage corresponds to the fermentation of a sugar contained in flour called maltose. Maltose comes from the action of some enzymes, the amylases, on the starch granules of the flour, damaged during the milling process. Amylases which are naturally present in flour, split starch into small fractions of a much simpler sugar, the maltose. The action of amylases starts as soon as water is added to the flour and stops during baking.
- The action of the flour amylases is completed by that of another enzyme of yeast, the maltase which, in its turn, splits maltose to give the most simple sugar, glucose. The glucose is transformed by the yeast into carbon dioxide and alcohol.
- Maltose formed from starch must be present in sufficient quantity so that the production of carbon dioxide makes dough rise correctly until it is put in the oven.
- Since 1960, the bakery industry has benefited from better adapted yeast strains, allowing a rapid process as they ferment the maltose sugar earlier. Effectively, the maltose is fermented only when there are no or little pre-existing sugars or added simple sugars left in the dough.
- The yeast industry has studied to a great extent baker's yeast strains, not only with a view to making them more rapid, but also to adapt them as much as possible to different types of baking processes.
- If sucrose has been incorporated into the dough, it is immediately transformed by a yeast enzyme, the invertase, into glucose and fructose (the two basic links composing sucrose are thus separated).
- When sucrose or glucose are added to the dough, they are directly fermented before maltose. This means that in such a dough, mainly sucrose or glucose are consumed by the yeast. These sugars do not contribute to giving the bread a sweet taste since they are partly consumed.

### 5.19.3 Yeast's Crucial Roles in Breadbaking

Yeast is the driving force behind fermentation, the magical process that allows a dense mass of dough to become a well-risen loaf of bread. And yet yeast is nothing more than a single-celled fungus. How does it do it?

Yeast works by consuming sugar and excreting carbon dioxide and alcohol as byproducts. In bread making, yeast has three major roles. Most of us are familiar with yeast's leavening ability. But you may not be aware that fermentation helps to strengthen and develop gluten in dough and also contributes to incredible flavors in bread.

### 5.19.4 Yeast Makes Dough Rise

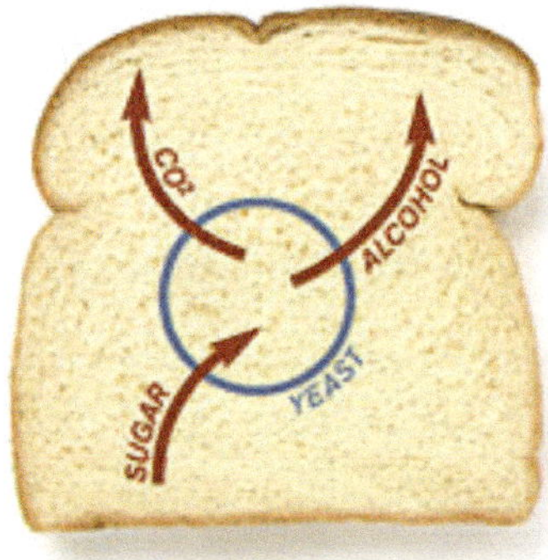


**Fig. 9. Yeast cells thrive on simple sugars. As the sugars are metabolized, carbon dioxide and alcohol are released into the bread dough, making it rise**

The essentials of any bread dough are flour, water, and of course yeast. As soon as these ingredients are stirred together, enzymes in the yeast and the flour cause large starch molecules to break down into simple sugars. The yeast metabolizes these simple sugars and exudes a liquid that releases carbon dioxide and ethyl alcohol into existing air bubbles in the dough.

If the dough has a strong and elastic gluten network, the carbon dioxide is held within the bubble and will begin to inflate it, just like someone blowing up bubblegum. As more and more tiny air cells fill with carbon dioxide, the dough rises and we're on the way to leavened bread.

### 5.19.5 Yeast Strengthens Bread Dough

When you stir together flour and water, two proteins in the flour—glutenin and gliadin—grab water and each other to form a bubblegum-like, elastic mass of molecules that we call gluten. In bread making, we want to develop as much gluten as we can because it strengthens the dough and holds in gases that will make the bread rise.

Once flour and water are mixed together, any further working of the dough encourages more gluten to form. Manipulating the dough in any way allows more proteins and water to find each other and link together. If you've ever made homemade pasta, you know that each time you roll the dough through the machine, the dough becomes more elastic; in other words, more gluten is developed. And with puff pastry dough, every time you fold, turn, and roll the dough, it becomes more elastic.

**Yeast, like kneading, helps develop the gluten network.** With every burst of carbon dioxide that the yeast releases into an air bubble, protein and water molecules move about and have another chance to connect and form more gluten. In this way, a dough's rising is an almost molecule-by-molecule kneading. Next time you punch down bread dough after its first rise, notice how smooth and strong the gluten has become, in part from the rise.

At this stage, most bakers stretch and tuck the dough into a round to give it a smooth, tight top that will trap the gases produced by fermentation. Then they let this very springy dough stand for 10 to 15 minutes. This lets the gluten bonds relax a little and makes the final shaping of the dough easier. This rounding and resting step isn't included in many home baking recipes, but it's a good thing to do.

### 5.19.6 Fermentation Generates Flavor in Bread

**Fig. 10. This loaf of artisan bread owes its complex flavor to a lengthy fermentation, which breaks down big molecules into smaller flavorful ones.**

As Harold McGee, the author of On Food & Cooking, has pointed out, big molecules in proteins, starches, and fats don't have much flavor, but when they break down into their building blocks—proteins into amino acids, starches into sugars, or fats into free fatty acids—they all have marvelous flavors. Fermentation, whether it's acting on fruit juices to make wine or on flour to make bread, does exactly that—it breaks down large molecules into smaller, flavorful ones.

At the beginning of fermentation, enzymes in the yeast start breaking down starch into more flavorful sugars. The yeast uses these sugars, as well as sugars already present in the dough, and produces not only carbon dioxide and alcohol but also a host of flavorful byproducts such as organic acids and amino acids. A multitude of enzymes encourages all kinds of reactions that break big chains of molecules into smaller ones—amylose and maltose into glucose, proteins into amino acids.

As fermentation proceeds, the dough becomes more acidic. This is due in part to rising levels of carbon dioxide, but there are also more flavorful organic acids like acetic acid (vinegar) and lactic acid being formed from the alcohol in the dough. The acidity of the dough causes more molecules to break down. The dough becomes a veritable ferment of reactions. Eventually, the amount of alcohol formed starts to inhibit the yeast's activity.

Yeast has help in producing flavorful compounds. Bacteria are important flavor builders as well. There are bacteria in the dough from the beginning, but as long as the yeast is very active, it consumes sugars as quickly as they're produced, leaving no food for the bacteria, which also like sugar. But when bakers chill a dough and slow down its rise, the cold dramatically reduces yeast activity. The bacteria, on the other hand, function well even in cold temperatures, so they now have an opportunity to thrive, producing many more marvelously flavorful acids.

## 5.20 BAKER'S YEAST PRODUCTION PRINCIPLES

Yeasts can grow in the presence or absence of air. Anaerobic growth, growth in the absence of oxygen, is quite slow and inefficient. For instance, in bread dough, yeast grow very little. Instead, the sugar that can sustain either fermentation or growth is used mainly to produce alcohol and carbon dioxide. Only a small portion of the sugar is used for cell maintenance and growth. In contrast, under aerobic conditions, in the presence of a sufficient quantity of dissolved oxygen, yeast grow by using most of the available sugar for growth and producing only negligible quantities of alcohol.

This means that the baker who is interested in the leavening action of carbon dioxide works under conditions that minimize the presence of dissolved oxygen. On the other hand, a yeast manufacturer that wants to produce more yeast cell mass, works under aerobic conditions by bubbling air through the solution in which the yeast is grown.

The problem posed to the yeast manufacturer, however, is not as simple as just adding air during the fermentation process. If the concentration of sugar in the fermentation growth media is greater than a very small amount, the yeast will produce some alcohol even if the supply of oxygen is adequate or even in abundance. This problem can be solved by adding the sugar solution slowly to the yeast throughout the fermentation process. The rate of addition of the sugar solution must be such that the yeast uses the sugar fast enough so that the sugar concentration at any one time is practically zero. This type of fermentation is referred to as a fed-batch fermentation.

**Fig. 11. Yeast Production**

### 5.20.1 Baker's Yeast Production Practice

The baker's yeast production process flow chart attached below can be divided into four basic steps. In order these steps are, molasses and other raw material preparation, culture or seed yeast preparation, fermentation and harvesting and filtration and packaging. The process outlined in the flow chart takes approximately five days from start to finish.

The basic carbon and energy source for yeast growth are sugars. Starch can not be used because yeast does not contain the appropriate enzymes to hydrolyze this substrate to fermentable sugars. Beet and cane molasses are commonly used as raw material because the sugars present in molasses, a mixture of sucrose, fructose and glucose, are readily fermentable. In addition to sugar, yeast also require certain minerals, vitamins and salts for growth. Some of these can be added to the blend of beet and cane molasses prior to flash sterilization while others are fed separately to the fermentation. Alternatively, a separate nutrient feed tank can be used to mix and deliver some of the necessary vitamins and minerals. Required nitrogen is supplied in the form of ammonia and phosphate is supplied in the form of phosphoric acid. Each of these nutrients is fed separately to the fermentation to permit better pH control of the process. The sterilized molasses, commonly referred to as mash or wort, is stored in a separate stainless steel tank. The mash stored in this tank is then used to feed sugar and other nutrients to the appropriate fermentation vessels.

Baker's yeast production starts with a pure culture tube or frozen vial of the appropriate yeast strain. This yeast serves as the inoculum for the pre-pure culture tank, a small pressure vessel where seed is grown in medium under strict sterile conditions. Following growth, the contents of this vessel are transferred to a larger pure culture fermentor where propagation is carried out with some aeration, again under sterile conditions. These early stages are conducted as set-batch fermentations. In a set-batch fermentation all the growth media and nutrients are introduced to the tank prior to inoculation.

From the pure culture vessel, the grown cells are transferred to a series of progressively larger seed and semi-seed fermentors. These later stages are conducted as fed-batch fermentations. During a fed-batch fermentation, molasses, phosphoric acid, ammonia and minerals are fed to the yeast at a controlled rate. This rate is designed to feed just enough sugar and nutrients to the yeast to maximize multiplication and prevent the production of alcohol. In addition, these fed-batch fermentations are not completely sterile. It is not economical to use pressurized tanks to guarantee sterility of the large volumes of air required in these fermentors or to achieve sterile conditions during all the transfers through the many pipes, pumps and centrifuges. Extensive cleaning of the equipment, steaming of pipes and tanks and filtering of the air is practiced to insure as aseptic conditions as possible.

At the end of the semi-seed fermentation, the contents of the vessel are pumped to a series of separators that separate the yeast from the spent molasses. The yeast is then washed with cold water and pumped to a semi-seed yeast storage tank where the yeast cream is held at 34 degrees Fahrenheit until it is used to inoculate the commercial fermentation tanks. These commercial fermentors are the final step in the fermentation process and are often referred to as the final or trade fermentation.

Commercial fermentations are carried out in large fermentors with working volumes up to 50,000 gallons. To start the commercial fermentation, a volume of water, referred to as set water, is pumped into the fermentor. Next, in a process referred to as pitching, semi-seed yeast from the storage tank is transferred into the fermentor. Following addition of the seed yeast, aeration, cooling and nutrient additions are started to begin the 15-20 hour fermentation. At the start of the fermentation, the liquid seed yeast and additional water may occupy only about one-third to one-half of the fermentor volume. Constant additions of nutrients during the course of fermentation bring the fermentor to its final volume. The rate of nutrient addition increases throughout the fermentation because more nutrients have to be supplied to support growth of the increasing cell population. The number of yeast cells increase about five- to eight-fold during this fermentation.

Air is provided to the fermentor through a series of perforated tubes located at the bottom of the vessel. The rate of airflow is about one volume of air per fermentor volume per minute. A large amount of heat is generated during yeast growth and cooling is accomplished by internal cooling coils or by pumping the fermentation liquid, also known as broth, through an external heat exchanger. The addition of nutrients and regulation of pH, temperature and airflow are carefully monitored and controlled by computer systems during the entire production process. Throughout the fermentation, the temperature is kept at approximately 86 degrees Fahrenheit and the pH in the range of 4.5-5.5.

At the end of fermentation, the fermentor broth is separated by nozzle-type centrifuges, washed with water and re-centrifuged to yield a yeast cream with a solids concentration of approximately 18%. The yeast cream is cooled to about 45 degrees Fahrenheit and stored in a separate, refrigerated stainless steel cream tank. Cream yeast can be loaded directly into tanker trucks and delivered to customers equipped with an appropriate cream yeast handling system. Alternatively, the yeast cream can be pumped to a plate and frame filter press and dewatered to a cake-like consistency with a 30-32% yeast solids content. This press cake yeast is crumbled into pieces and packed into 50-pound bags that are stacked on a pallet. The yeast heats up during the pressing and packaging operations and the bags of crumbled yeast must be cooled in a refrigerator for a period of time with adequate ventilation and placement of pallets to permit free access to the cooling air. Palletized bags of crumbled yeast are then dis tributed to customers in refrigerated trucks.

# 6

# INDUSTRIAL FOOD FERMENTATION

## 6.1 INTRODUCTION

Fermentation is the process of converting carbohydrates to alcohol or to organic acids using microorganisms, such as yeasts or bacteria. This process takes place when there are beneficial bacteria present that break down the starch and sugars in the food. As the microorganisms divide, lactic acid is formed, which stops the growth of bad bacteria. The lactic acid is also what gives fermented foods that very specific 'tangy' or 'acidic' taste. Fermented foods can last for many months (some products even lasting multiple years) as long as they are stored in a cool, dark place and kept in the solution of salt and water known as brine.

For thousands of years, before the invention of refrigerators and freezers, the most common way to preserve food from spoiling was through fermentation. Fermentation has existed since the Neolithic Era, some of the earliest documentation of fermentation being between 7000-6600 BCE in Jiahu, China.

Fermentation can also make food healthier and safer to consume. For instance, in the Middle Ages, drinking water was hazardous because it frequently contained pathogens that could spread disease. Making the water into beer made it safe to drink, as any deadly bacteria previously present in the water were killed during the brewing process. Additionally, since microorganisms can produce vitamins as they ferment, the beer had added nutrients in it from the barley and various other ingredients.

In today's Western society, we don't eat quite as many fermented foods as our ancestors did. The most commonly consumed fermented food and drink would be cheese, beer, wine, yogurt, cured sausage and sourdough bread. Many store-bought fermented foods, like sauerkraut or pickles, are not authentic as they are often preserved in vinegar instead of the traditional and naturally occurring beneficial bacteria. Due to North America's strict health and safety concerns, these foods are also often pasteurised, which robs the foods of their nutrients and minerals.

Industrial fermentation is that the intentional use of fermentation by microorganisms like microorganism and fungi to form product helpful to humans. soured product have applications as food still as generally trade. Some trade goods chemicals, like carboxylic acid, acid, and grain alcohol square measure created by fermentation. The speed of fermentation depends on the concentration of microorganisms, cells, cellular elements, and enzymes still as temperature, pH. And for aerobic fermentation element. Product recovery oftentimes involves the concentration of the dilute resolution. Nearly all commercially made enzymes, like enzyme, disaccharidase and organic compound, square measure created by fermentation with genetically changed microbes. In some cases, production of biomass itself is that the objective, as within the case of Saccharomyces cerevisiae and carboxylic acid microorganism starter cultures for cheesemaking. In general, fermentations is divided into four types:

- Production of biomass (viable cellular material)
- Production of extracellular metabolites (chemical compounds)
- Production of animate thing elements (enzymes and different proteins)
- Transformation of substrate (in that the reworked substrate is itself the product),

These sorts aren't essentially disjoint from one another; however offer a framework for understanding the variations in approach. The organisms used are also microorganism, yeasts, molds, animal cells, or plant cells. Special issues square measure needed for the particular organisms utilized in the fermentation, like the dissolved element level, nutrient levels, and temperature.

## 6.2 GENERAL PROCESS

In most industrial fermentations, the organism's area unit submerged in an exceedingly liquid medium; in others, like the fermentation of cocoa beans, occasional cherries, and miso, fermentation takes place on the damp surface of the medium. There are industrial concerns associated with the fermentation method. as an example, to avoid organic process contamination, the fermentation medium, air, and instrumentation area unit sterilized. Foam management will be achieved by either mechanical foam destruction or chemical anti-foaming agents. Many alternative factors should be measured and controlled like pressure, temperature, troublemaker shaft power, and consistency.

A vital component for industrial fermentations is rescaling. This can be the conversion of a laboratory procedure to Associate in Nursing process. it's well established within the field of commercial biology that what works well at the laboratory scale may go poorly or not the least bit once initial tried at massive scale. It's usually uphill to require fermentation conditions that have worked within the laboratory and blindly apply them to industrial-scale instrumentation. though several parameters are tested to be used as rescale criteria, there's no general formula attributable to the variation in fermentation processes. The foremost vital strategies area unit the upkeep of constant power consumption per unit of broth and also the maintenance of constant volumetrically transfer rate.

## 6.2.1 Microbial Growth

When a selected organism is introduced into a specific growth medium, the medium is inoculated with the actual organism. Growth of the substance doesn't occur directly, however takes a bit whereas. this can be the amount of adaptation, known as the lag part. Following the lag part, the speed of growth of the organism steady will increase, for a precise amount—this period is that the log or exponential part.After a precise time of exponential part, the speed of growth slows down, thanks to the unceasingly falling concentrations of nutrients and/or a unceasingly increasing (accumulating) concentrations of toxic substances. This phase, wherever the rise of the speed of growth is checked, is that the slowing part. once the slowing part, growth ceases and therefore the culture enters a stationary part or a gentle state. The biomass remains constant, except once sure accumulated chemicals within the culture lyse the cells (chemolysis).

Unless alternative micro-organisms contaminate the culture, the chemical constitution remains unchanged. If all of the nutrients within the medium square measure consumed, or if the concentration of poisons is just too nice, the cells might become scenescent and start to die out. the entire quantity of biomass might not decrease, however the amount of viable organisms can decrease.

## 7.2.2 Fermentation Medium

The microbes used for fermentation grow in (or on) specially designed growth medium that provides the nutrients needed by the organisms. a range of media exist, however invariably contain an carbon supply, a atomic number 7 supply, water, salts, and micronutrients. within the production of wine, the medium is grape should, within the production of bio-ethanol, the medium could consist principally of no matter cheap carbon supply is obtainable, Carbon sources square measure usually sugars or different carbohydrates, though within the case of substrate transformations (such because the production of vinegar) the carbon supply is also associate degree alcohol or one thing else altogether. for giant scale fermentations, like those used for the assembly of plant product, cheap sources of carbohydrates, like syrup, corn steep liquor, sugar cane juice, or sugar beet juice square measure wont to minimize prices. Additional sensitive fermentations could instead use sublimate aldohexose, sucrose, glycerin or different sugars, that reduces variation and helps make sure the purity of the ultimate product.

Organisms meant to supply enzymes like beta galactosidase, saccharase or different amylases are also fed starch to pick for organisms that categorical the enzymes in lots of abundance. Fixed atomic number 7 sources square measure needed for many organisms to synthesize proteins, nucleic acids and different cellular elements. Reckoning on the protein capabilities of the organism, atomic number 7 is also provided as bulk macromolecule, like soy meal; as pre-digested polypeptides, like organic compound or tryptone; or as ammonia or nitrate salts.price is additionally a very important think about the selection of a atomic number 7 supply. Phosphorus is required for production of phospholipids in cellular membranes and for the assembly of nucleic acids. The quantity of phosphate that should be additional depends upon the composition of the broth and also the wants of the organism, similarly because the objective of the fermentation. For example, some cultures won't turn out secondary metabolites within the presence of phosphate. Growth factors and trace nutrients square measure enclosed within the fermentation broth for organisms incapable of manufacturing all of the vitamins they need. Yeast extract could be a common supply of micronutrients and vitamins for fermentation media. Inorganic nutrients, as well as trace parts like iron, zinc, copper, manganese, atomic number 42 and metal square measure usually gift in unrefined carbon and atomic number 7 sources, however could have to be compelled to be additional once sublimate carbon and atomic number 7 sources square measure used. Fermentations that turn out giant amounts of gas (or that need the addition of gas) can tend to make a layer of froth, since fermentation broth usually contains a range of foam-reinforcing proteins, peptides or starches. to forestall this foam from occurring or accumulating, antifoaming agents is also additional.

### 6.2.3 Fermenters

Industrial fermentations ar usually administered in giant tanks, known as fermenters or bioreactor. reckoning on the character of the fermentation, gas could also be sparged into the fermentation medium. For aerobic fermentations, air is usually used as a result of its affordable thanks to give enough gas for metastasis. Anaerobic fermentations, like the assembly of fermentation alcohol, usually don't need the addition of any air, and solely need agitation from a mixer to stay the organisms suspended. Aerobic fermentations could also be conducted in a very form of fermenters, like a bubble column or a packed bed over those fermentation medium drips. Cooling is usually needed, since organisms turn out waste heat as a part of their metabolism.

## 6.3 PRODUCTION OF INTRACELLULAR COMPONENTS

Production of intracellular elements of primary interest among the intracellular elements area unit microorganism enzymes: enzyme, amylase, protease, pectinase, aldohexose enzyme, cellulase, hemicellulase, lipase,lactase, enzyme and lots of others. Recombinant proteins, like hypoglycaemic agent, serum hepatitis immunogen, interferon, white blood corpuscle colony-stimulating issue, enzyme et al. are created this fashion. The most important distinction between this method and also the others is that the cells should be burst (lysed) at the top of fermentation, and also the surroundings should be manipulated to maximise the number of the merchandise. What is more, the merchandise (typically a protein) should be separated from all of the opposite cellular proteins within the lysate to be pure.

### 6.3.1 Transformation of Substrate

Substrate transformation involves the transformation of a particular compound into another, like within the case of phenylacetylcarbinol, and steroid biotransformation, or the transformation of a stuff into a finished product, within the case of food fermentations and waste product treatment.

## 6.4 FOOD FERMENTATION

Ancient soured food processes, like creating bread, wine, cheese, curds, idli, dosa, etc., are often dated to quite seven thousand years agone. They were developed long before man had any data of the existence of the microorganisms concerned. Some foods like Marmite area unit the byproduct of the fermentation method, during this case within the production of brew.

### 6.4.1 Ethanol Fuel

Fermentation is that the main supply of ethyl alcohol within the production of ethyl alcohol fuel. Common crops like sugar cane, potato, cassava and corn area unit soured by yeast to provide ethyl alcohol that is more processed to become fuel.

## 6.5 TRADITIONAL FERMENTATION TECHNOLOGY

Fermentation technology is more than 3000 year old in India. The fermentation technology employed a variety of processes and was put to a large number of uses. It also laid the foundation of alchemy and chemistry.

Fermentation is the process of digesting certain substances that leads to chemical conversion of organic substances into simpler compounds. Fermentation technology began with sweet substances (vegetable or animal) in different parts of the world. The process of fermentation was probably discovered by observing the changes in the juices of several fruits and other substances that had been kept for a day or more. It appears that fermentation technology started simultaneously with settled agriculture during the Neolithic period.

A sweet substance known as Soma juice prepared by the Vedic Aryans is supposed to be the first product of fermentation in India. The Rgveda (c.1500 BC) shows that fermentation technology took its first step in connection with the preparation of Soma juice in India. There is also another drink, known as Sura (wine/beer), prepared by fermentation. These two preparations have also been used in different medicinal preparations, surgical procedures and in many chemical and alchemical operations. It is believed that acetic fermentation was known to India since the early times.

### 6.5.1 Vedic Period

The two drinks that are quoted in Vedic texts are Soma and Sura. The Rgveda describes soma as a godly concoction, while sura is described as mainly a human drink. However, the Yajurveda deifies sura and uplifts it to the status of soma.

### 7.5.2 Evidence of Fermentation in the Rigveda

Soma is an invigorating plant product generally prepared by squeezing juice from plants. The following statements give an indication of fermentation:

- Admixture of a thick juice of Soma with barley powder (RV. IX. 68.4). This statement is a clear indication of fermentation with barley.
- "Fifteenth day old highly intoxicating Soma" (RV. X.27.2). In this statement fifteenth day probably refers to the broth fermented in the vat for 15 days.
- "Soma being treated is red" (RV.IX.82: 1).
- All these statements give an indication of fermentation technology involved in the preparation of Soma.
- Evidence of fermentation in the Yajurveda.
- The Sukla Yajurveda (Vaj. XIX. 13-15; 82-83) describes the formation of two stimulating drinks – Sura andParisrut. Sura was supposed to be prepared from germinated paddy, germinated barley, and parched rice with the help of ferment. Yeast was used most often as the fermenting agent.
- The Katyayana Srauta sutra (XV, 9.28-30; XIX, 1-2) also gives a complete description of the preparation of Sura. According to this method either boiled rice or boiled barley was mixed with the ferment and also with masara and the entire mixture was kept in a jar. The jar was then kept in a pit for three nights. Meanwhile, cow's milk and powdered parched rice were poured in the pit.

### 7.5.3 Coagulation of Milk

A large number of dairy products that used fermentation technology are mentioned in the Rigveda. Dahi (curd) was such a product that was used as an essential offering to God and was also an ingredient for the preparation of somajuice. Different methods of curdling of milk are described in different texts. For instance, the method described in Taittiriya Samhita mentions the use of whey and rice grains for curdling of milk.

## 6.6 POST-VEDIC PERIOD (600 BC TO 100 CE)

In this period, two new preparations known as vinegar and liquor from bread were added by adopting new technologies. Later on some other drinks were also prepared by using fermentation technologies. Various developments took place in the field of fermentation technology during the post-Vedic period. Some of these developments are described below:

- Now, a large number of base materials were used instead of a few. Some common substances that began to be used as base material were rice and barley among cereals; grape, palm, mango, wood-apple and sugarcane, etc, among fruits; madhuka and jasmine among flowers and barks ofmesasrngi, putraka, etc.

- Earlier honey was mostly used as a ferment, but later on treacle (AS. II.25) also began to be used and became very popular in a short span of time.
- A new period began with the employment of fermented products in therapeutic medicines and chemical practices.
- Auxiliary (sambhara) (AS.II.25): It was a compound that served dual purposes. On the one hand it was used as ferment, on the other it was also used to add flavour and taste to liquor. This compound was prepared by using cinnamon, Plumbago zeylanica, Emblia ribes and gaza-pippali (elephant pepper plant). From sambhara a special liquid known as Sambhariki was also prepared.

### 6.6.1 Technologies Connected with the Preparation of Beer (Sura)

The Buddhist texts mention five different types of sura, but the technologies involved in the preparation of these are not discussed. Some texts mention that germinated cereals and hulled fruit drenched in water dried, ground and given the shape of a cake was known as bread. Later on brewing was done by soaking it in water which in turn was fermented by adding some yeast to that solution.

## 6.7 TECHNOLOGIES CONNECTED WITH THE PREPARATION OF VINEGAR (AMLA)

According to Kautilya, there were two types of amla preparations. One method involved the use of fruit juice and phalamla, while the other was derived from molasses. Its preparation was done for a longer time so that the product became sour. Kautilya further states that these were home-made products.

Medaka (spiced rice beer), Prasanna (spiced barley or wheat beer), Asava (sugarcane beer), Arista (medical tincture), etc, were some of the most popular drinks of that period. Different varieties of fermented liquor provide us the knowledge of different fermentation processes developed during that period.

## 6.8 PERIOD OF SCIENTIFIC AND TECHNICAL LITERATURE (100 CE ONWARDS)

Fermentation technology advanced in the field of medical science and chemical and alchemical practices. A large number of digesting methods now came into existence. Some of these digesting methods were:

- The fermenting vessel was deposited into the earth and sometimes the vessel was covered with horse dung.
- The fermenting vessel was placed on a heap of grains.
- The fermenting vessel was exposed to the sun.
- Fumigation of the desired substance was done in a pot.

Some new recipes for fermentation were formulated. One such recipe is described in Rasopanisat (RP. XV 251-253). The recipe is:

"The best ones among the five classes of bulbous plant with latex pounded along with the grains of rice of Kodrava (Paspaln scrobiculatum), and products of plant madana (emetic nut), pasted with whey of buffalo or cow-milk and kept in a closed bowl and another. The closed vessel is then placed in the sun. The acidic residue thus obtained iskinva (yeast)."

The alcoholic fermentation and acetous fermentation began to be used for several purposes now by changing the base material or by slightly modifying the digesting methods. Base material was known as yoni and the fermented produces were known as tri-yoni, dvi-yoni, etc. The juicy produce that was obtained from the fermented smash was known as Jata-rasa. The liquid obtained from the boiled substance was also used for the processes of fermentation.

Fermentation products, arista, asava and sura, were considered potent medicinal drugs. Similarly, aranala, cukra, dhanyamala, kanjika, sauviraka, sukta, tusambu, etc, were not only used for medicinal purposes but also used for washing of metals and minerals and for different mercurial operations.

## 6.9 PRACTICES INVOLVED IN SPIRITUOUS (ALCOHOLIC) FERMENTATION

***Arista :*** Powder of certain selected drugs of desired medicinal properties as well as fermenting properties mixed with water, iron powder and treacle was used for the preparation of arista. Water taken was about half of the weight of the powder while iron powder and treacle taken were 28 ½ times the weight of water. All of these ingredients were placed in an earthen vessel. The earthen vessel was coated with a mixture of honey, butter and powdered long-pepper plant from inside. Now the sealed earthen pot was placed in a mass of barley for at least seven nights for fermentation to take place.

***Asava :*** A mixture of either cold alkaline water or sesamum and juice of sugarcane was taken for the preparation ofasava. These were taken in the ratio of 2:3 respectively. The process of fermentation was the same as that ofarista.

***Sura :*** In the preparation of Sura a decoction of a mixture of desirable drugs was taken with the ferment. This preparation was also similar to that of arista and asava.

### 6.9.1 Preparation of Special Sura for Purification

A mixture of decoctions of prescribed plants and sugar was taken in the ratio of 2: 3 respectively. This mixture was then boiled over fire. After boiling the mixture, the solution was kept for cooling. Now, the solution was placed in an earthen jar for about a month. The time period was taken in such a way that half of this period fell in the winter and the other half in the anterior or posterior of winter. The fermentation process took place during this period. Likewise, alkaline water, cow's urine and some other substances were also used for similar preparations.

Not only references to these medicated liquors but a number of other fermented liquors with their respective medicinal values are found in Caraka and Susruta-Samhita.

### 7.9.2 Practices Involved in Acetic Fermentation

Some common practices are described below:

**Aranala:** Husked uncooked wheat or any cereal was taken as a base material with any ferment and the mixture was heated for about a week by solar radiation.

**Cukra:** Bulging root, fruit of the desired plant was taken with alcoholic sweet juicy whey and the mixture was deposited and purified in a heap of paddy for about three nights.

**Sukta:** A mixture of treacle, honey, fermented rice water, whey was used in the preparation of sukta. This mixture was placed in an earthen pot and the pot was placed on heaps of paddy for three nights in the summer season.

**Kanjika:** Boiled millet or barley was used as a base material. Different plants were used as a ferment medium. It could also be prepared by using boiled rice.

The above facts clearly show that the history of fermentation technology in India from the ancient times onwards includes a variety of techniques which resulted in a rich repertoire of alcoholic beverages and allied products. What is important is that fire and heat were rarely employed for the process of fermentation.

## 6.10 BENEFITS OF TRADITIONAL FERMENTED FOOD

All around the world, fermented foods and beverages are part of the human diet. In some places they make up a minor 5% of daily intake, while in others their role can be as substantial as 40%.

Using native knowledge of locally available raw materials from plant or animal sources, people across the globe produce this type of food and drink either naturally or by adding starter cultures that contain micro-organisms. Micro-organisms transform these raw materials both biochemically (i.e., the nutrients) andorganoleptically (i.e., the taste/texture/odour) into edible products that are culturally acceptable to the maker and consumer.

Fermented foods can be fried, boiled or candied, or consumed in curries, stews, side dishes, pickles, confectionery, salads, soups and desserts. They can be in the form of pastes, seasonings, condiments, masticators, and even colourants. Fermented drinks can be either alcoholic (such as beer and wine) or non-alcoholic, like butter milk, certain teas, or things that contain vinegar.

However, though most fermented foods have health-promoting benefits, their global consumption is declining as traditional food systems give way to the influence of a western diet and fast foods.

### 6.10.1 A World of Therapeutic Food

Many of the fermented products consumed by different ethnic groups have therapeutic values. Some of the most widely known are fermented milks (i.e., yoghurt, curds). Containing high concentrations of pro-biotic bacteria, these can lower your cholesterol level.

**Fig. 1. Kefir is easily digested and provides beneficial micro-organisms that contribute to a healthy immune system.**

One fermented food that has received much attention is the viscous, acidic, mildly alcoholic milk beverage produced by fermentation of milk with a particular grain in Eastern European and Middle Eastern countries. Kefir is easily digested and provides the human body with beneficial micro-organisms that contribute to a healthy immune system — a boon to patients suffering from AIDS, chronic fatigue syndrome, herpes and cancer. Traditionally it has been used for the treatment of tuberculosis and cancer. Another naturally fermented dairy product, koumiss, found in the Caucasus region, is used in the treatment of pulmonary tuberculosis.

**Fig. 2. Antioxidant and other properties are reported in foods such as natto, a sticky dish high in protein and popular at breakfast in Japan.**

Other widely-eaten types of health-boosting sustenance include fermented soybean products found around Asia. Antioxidant and other properties are reported to exist in foods such as natto, a sticky dish high in protein and popular at breakfast in Japan, that may help prevent people from having brain haemorrhages. Natto is also rich in vitamin K2, which stimulates the formation of bones and might help to prevent osteoporosis in older people. Similarly, consumption of Indonesia's tempe reduces cholesterol levels and, like China's douchi, lowers high blood pressure.

**Fig. 3. China's douchi lowers high blood pressure.**

Korea's famous accompaniment, kimchi, may take the prize for most benefits however, reportedly helping to prevent constipation and colon cancer and reduce serum cholesterol, as well as possessing anti-stress effects and the ability to ameliorate depression, osteoarthritis, liver disease, obesity and atherosclerosis.

In the Himalayas, a fermented leafy vegetable product called gundruk and a fermented radish tap-root have large amounts of lactic acids, ascorbic acid, carotene and dietary fibre, which have anti-carcinogenic effects. In order to gain strength, ailing persons and post-natal women in the Himalayas consume bhaati jaanr extract and kodo ko jaanr due to their high calorie content.

## 6.11 FERMENTATION'S MAGIC

One of the important outcomes of food fermentation is the enrichment of food with essential amino acids, vitamins, mineral and bioactive compounds. For example, during tempe fermentation, vitamins like niacin and riboflavin are increased by using the starter culture Rhizopus oligosporus. Similarly, thiamine and riboflavin increase during fermentation of idli, the fermented rice and black-gram (a type of legume) product of India and Sri Lanka.

**Fig. 4. Pulque is one of the oldest alcoholic beverages and is prepared from the juices of cactus plants in Mexico.**

These clever processes developed by humans to extract additional nutritional value from food, are not only found in Asia. Pulque — one of the oldest alcoholic beverages prepared from the juices of cactus plants in Mexico — is rich in vitamins such as thiamine, riboflavin, niacin and biotin.

Another important reason to ferment foods is to coax micro-organisms into producing enzymes that also provide very useful services. These enzymes degrade anti-nutritive compounds and thereby make edible, with enhanced flavour and aroma, things that otherwise would be indigestible and/or unpalatable. Bitter varieties of cassava tubers contain a potentially poisonous substance that can be detoxified via lactic acid bacteria, as in gari and fufu, fermented cassava root products from Africa.

**Fig. 5. Bitter varieties of cassava can be detoxified via lactic acid bacteria, as are gari and fufu in Africa.**

Many people suffer from lactose intolerance or lactose malabsorption, a condition that causes lactose, the principal carbohydrate of milk, to not be completely digested. The cultures used in making yoghurt and curds, contain substantial quantities of ß-D-galactosidase, something that is thought to help alleviate the symptoms of lactose malabsorption. Yoghurt, as a viscous food, may delay the stomach emptying and that way help lessen lactose intolerant symptoms.

### 6.11.1 Healthy, but Not Appreciated

Although fermented foods are marketed globally as health foods, functional foods, therapeutic foods, nutraceutical foods or bio-foods, due to urbanization, changes in life-style, and the shifting from traditional food habits to commercial fast foods, the production and consumption of traditional fermented foods is in decline.

Even when fermented foods are being consumed, they are increasingly likely to be from packets in supermarkets, rather than directly from households or a local family run business. This is leading to fewer and fewer people possessing traditional knowledge of fermented foods and the ability to make these from scratch. Reliance on fewer providers of fermented foods is also leading to a decline in the biodiversity of micro-organisms, or 'microbial-biodiversity'.

## 6.12 IMPROVING THE EFFICIENCY OF TRADITIONAL FERMENTATION INDUSTRIES

Most of the traditional food fermentation industries could be considered as small household or backyard industries with small capacity production and inefficient in the application of the fermentation technology process and know how. It is not that difficult to improve or upgrade the traditional fermentation industries with little capital input and application of GMP,

The characteristics of traditional fermentation industries in Malaysia are:

1. They are small scale or small volume industries
2. There are a diversity of fermentation products produced mre towards the geographical location and the type of food substrate available in the area. Coastal areas may produce fermentation products from cincaluk, belacan, budu. Other more inland areas might be producing tempe, tapai and kicap fermentation
3. Level of technology utilized is very low and do not really changed with time and regarded as passive fermentation. Use of automation and mechanization is low
4. GMP is rarely applied. Lack of quality control and sanitary standards
5. Very low level of research or scientific input

6. Government industries supposed to support and upgrade these small and scattered industries are quite passive and do not really enter the playground to participate
7. Shelf life for some of the fermentation products are really short in case of tempe and tapai
8. Marketting of the traditional fermentation products considered poor
9. These industries do not really have professional consultants in fermentation who can advise them how to improve and standardize the quality of fermentations
10. Most of the traditional fermentations are Batch or fed batch fermentations
11. Operators of the traditional fermentation have poor or lack of understanding of the science of fermentation that be applied to their process
12. The fermentation industry is a mix of Solid substrate and liquid fermentation

In this write up we will only try to discuss one or two aspects of the traditional fermentation industries

1. Increasing the volume of fermentation production
2. Improving the efficiency of the traditional fermentation industries with simple cheaper and alternative modes of fermentation

### 7.12.1 Increasing the Volume of Fermentation

One of the most common characteristics of traditional fermentation here is that the fermentation process are often carried out in a large number of small fermentation vats (or earthen jars). This limiting step can be avoided by carrying the fermentation process in a large fermentation vessel. The most ideal container for the fermentation now would be adapting or modifying the new stainless steel water tanks. These tanks are easily available from hardwate shops. These tanks are sanitary and are easily cleaned and sterilize.

Adding a simple motor to the top of the fermentor or stainless steel water tank will help improve the mixings and the fermentation process.

### 6.12.2 Increasing the Efficiency of the Fermentation Process

Improvement of the traditional fermentation process has always been the rate limiting steps in the fermentation process. To improve the fermentation process must be carried out in terms of time, quality and volume of the fermentation process.

When we talk about shortening the fermentation time by optimizing the various physical, chemical and microbiological as well as the engineering aspects of the fermentor operation. This initially requires some laboratory scale fermentation studies to optimize the various parameters

The product of the fermentation should be of the highest quality and consistency. Standard operating procedures and good GMP or the basics of GMP should be adapted or modified for use at the level of cottage industries fermentation. The HAACP procedures should also be adapted in its essence to be applied at the small scale fermentation

The main problem of traditional fermentation industries in this country it has always been so successful in maintaining the traditional way of thinking and making fermentations. There is lack of tendency to understand the process, to increase the input of science and technology to the fermentation production system.

In my observations the fermentation entrepreneurs still practise the art of fermentation handed down from the fathers and refusing to optimise or even experiment to improve their fermentation process. There is poor monitoring of the fermentation process in terms of temperature, ph, sugar in the traditional fermentation industries. This has led to inconsistency in the fermentation products produced. A little investment in these instruments would increase the efficiency of their fermentation.

## 6.13 ADVANCES IN TRADITIONAL FERMENTATION

Fermentation technology has been a long-favored biological process due to its simplicity of reaction, high specificity, low costs, and versatility of application. Modern industries have complemented the basic principle of fermentation with advances in techniques and genetic engineering to extend its application to the production of a range of biochemicals, biomolecules, and biofuels. Fermentation's applications are rapidly breaching the confines of food stuff such as wine, cheese, bread, and beer, into high-value chemicals, pharmaceutical ingredients, and food-related chemicals. Further, rising petrol prices and depleting fossil fuel reserves have made a robust case for low-cost and simple fermentation processes for producing drop-in chemical replacements. There has been a spurt in research efforts, both in the corporate and academic sector, driven largely by fermentation applications within the energy and utilities, specialty chemicals, and pharmaceuticals industries. Energy or more specifically, biofuels has evinced considerable interest in applying fermentative techniques to produce cost-competitive bioethanol.

While the uses of fermentation techniques are many in modern industrial processes, there are several challenges to overcome before implementing them in large-scale commercial applications. One of the stand-out challenges is that fermentation is a slow, natural bioprocess, which means that time-scale of operations is often quite long. Companies also have to contend with the low yields and poor process economics in large-scale production. Although fermentation operating costs are lower than full-scale chemical operations, the costs of raw materials and fermentable sugars remain a challenge for most companies. In addition, specially designed enzymes and microbes with specific biosynthetic pathways are also expensive. Therefore, the end products derived will be hard pressed to be cost competitive with other petrochemical-derived alternatives.

To stave off these challenges, companies are increasingly developing robust proprietary microorganisms that can ferment the substrates to produce the desired products. Consequently, companies that are able to provide a solution that combines excellence in microbe technologies with superiority in process technologies will be better equipped to explore new application avenues. Importantly, companies that can take advantage of fermentation's unique bio-capabilities, while still remaining economically feasible, will find themselves operating in a market with expansive possibilities. Many companies have begun engineering innovative enzymes, microbes, and bacterial strains that can help produce biomolecules in bulk. Apart from genetic techniques, some of the recent developments that continue to be explored include process-driven technologies such as solid-state fermentation, consolidated bioprocessing, ultrasonication, syngas fermentation, and dark fermentation. These developments are indicative of the huge potential of fermentation as a low-cost and flexible method to produce a range of end products.

### 7.13.1 Solid State Fermentation for the Production of Industrial Enzymes

Enzymes are among the most important products obtained for human needs through microbial sources. A large number of industrial processes in the areas of industrial, environmental and food biotechnology utilize enzymes at some stage or the other. Current developments in biotechnology are yielding new applications for enzymes. Solid state fermentation (SSF) holds tremendous potential for the production of enzymes. It can be of special interest in those processes where the crude fermented products may be used directly as enzyme sources. This review focuses on the production of various industrial enzymes by SSF processes. Following a brief discussion of the micro-organisms and the substrates used in SSF systems, and aspects of the design of fermenter and the factors affecting production of enzymes, an illustrative survey is presented on various individual groups of enzymes such as cellulolytic, pectinolytic, ligninolytic, amylolytic and lipolytic enzymes, etc.

Solid state fermentation (SSF) holds tremendous potential for the production of enzymes. It can be of special interest in those processes where the crude fermented product may be used directly as the enzyme source1. In addition to the conventional applications in food and fermentation industries, microbial enzymes have attained significant role in biotransformations involving organic solvent media, mainly for bioactive compounds.

## 6.14 MICRO-ORGANISMS USED FOR THE PRODUCTION OF ENZYMES IN SOLID STATE FERMENTATION SYSTEMS

A large number of microorganisms, including bacteria, yeast and fungi produce different groups of enzymes. Selection of a particular strain, however, remains a tedious task, especially when commercially competent enzyme yields are to be achieved. For example, it has been reported that while a strain of Aspergillus niger produced 19 types of enzymes, a-amylase was being produced by as many as 28 microbial cultures. Thus, the selection of a suitable strain for the required purpose depends upon a number of factors, in particular upon the nature of the substrate and environmental conditions. Generally, hydrolytic enzymes, e.g. cellulases, xylanases, pectinases, etc. are produced by fungal cultures, since such enzymes are used in nature by fungi for their growth.Trichoderma spp. and Aspergillus spp. have most widely been used for these enzymes. Amylolytic enzymes too are commonly produced by filamentous fungi and the preferred strains belong to the species of Aspergillus and Rhizopus. Although commercial production of amylases is carried out using both fungal and bacterial cultures, bacterial a -amylase is generally preferred for starch liquefaction due to its high temperature stability. In order to achieve high productivity with less production cost, apparently, genetically modified strains would hold the key to enzyme production.

The selection of a substrate for enzyme production in a SSF process depends upon several factors, mainly related with cost and availability of the substrate, and thus may involve screening of several agro-industrial residues. In a SSF process, the solid substrate not only supplies the nutrients to the microbial culture growing in it but also serves as an anchorage for the cells. The substrate that provides all the needed nutrients to the microorganisms growing in it should be considered as the ideal substrate. However, some of the nutrients may be available in sub-optimal concentrations, or even absent in the substrates. In such cases, it would become necessary to supplement them externally with these. It has also been a practice to pre-treat (chemically or mechanically) some of the substrates before using in SSF processes (e.g. ligno-cellulose), thereby making them more easily accessible for microbial growth.

Among the several factors that are important for microbial growth and enzyme production using a particular substrate, particle size and moisture level/water activity are the most critical. Generally, smaller substrate particles provide larger surface area for microbial attack and, thus, are a desirable factor. However, too small a substrate particle may result in substrate agumulation, which may interfere with microbial respiration/aeration, and therefore result in poor growth. In contrast, larger particles provide better respiration/aeration efficiency (due to increased inter-particle space), but provide limited surface for microbial attack. This necessitates a compromised particle size for a particular process.

SSF processes are distinct from submerged fermentation (SmF) culturing, since microbial growth and product formation occurs at or near the surface of the solid substrate particle having low moisture contents. Thus, it is crucial to provide an optimized water content, and control the water activity (aw) of the fermenting substrate—for, the availability of water in lower or higher concentrations affects microbial activity adversely. Moreover, water has profound impact on the physico-chemical properties of the solids and this, in turn, affects the overall process productivity.

### 6.14.1 Aspects of Design of Fermenter for Enzyme Production in Solid State Fermentation Systems

Over the years, different types of fermenters (bioreactors) have been employed for various purposes in SSF systems. Laboratory studies are generally carried out in Erlenmeyer flasks, beakers, petri dishes, roux bottles, jars and glass tubes (as column fermenter). Large-scale fermentation has been carried out in tray-, drum- or deep-trough type fermenters. The development of a simple and practical fermenter with automation, is yet to be achieved for the SSF processes.

### 6.14.2 Factors Affecting Enzyme Production in Solid State Fermentation Systems

The major factors that affect microbial synthesis of enzymes in a SSF system include: selection of a suitable substrate and microorganism; pre-treatment of the substrate; particle size (inter-particle space and surface area) of the substrate; water content and aw of the substrate; relative humidity; type and size of the inoculum; control of temperature of fermenting matter/removal of metabolic heat; period of cultivation; maintenance of uniformity in the environment of SSF system, and the gaseous atmos-phere, i.e. oxygen consumption rate and carbon dioxide evolution rate.

### 6.14.3 Enzymes Produced by Solid State Fermentation Processes

Ideally, almost all the known microbial enzymes can be produced under SSF systems. Literature survey reveals that much work has been carried out on the production of enzymes of industrial importance, like proteases, cellulases, ligninases, xylanases, pectinases, amylases, glucoamylases, etc.; and attempts are also being made to study SSF processes for the production of inulinases, phytases, tannases, phenolic acid esterases, microbial rennets, aryl-alcohol oxidases, oligosaccharide oxidases, tannin acyl hydrolase, a -L-arabinofuranosidase, etc. using SSF systems.

### 6.14.4 Cellulases, Xylanases and Xylosidases

Cellulases are a complex enzyme system, comprising endo-1,4-b -D-glucanase (EC-3.2.1.4), exo-1,4-b-glucanase (exocellobiohydrolase, EC-3.2.1.91) and b-D-glucosidase (b-D-glucoside glucanhydrolase, EC-3.2.1.21). These enzymes, together with other related enzymes, viz.hemicellulases and pectinases, are among the most important group of enzymes that are employed in the processing of ligno-cellulosic materials for the production of feed, fuel, and chemical feedstocks. Cellulases and xylanases (endo-1,4-b -D-xylanase, EC-3.2.1.8) however find applications in several other areas, like in textile industry for fibre treatment and in retting process. Xylanases find specific application in jute fibre upgradation also.

Currently, industrial demand for cellulases is being met by production methods using submerged fermentation (SmF) processes, employing generally genetically modified strains of Trichoderma. The cost of production in SmF systems is however high and it is uneconomical to use them in many of the aforesaid processes. This therefore necessitate reduction in production cost by deploying alternative methods, for example the SSF systems.

Tengerdy compared cellulase production in SmF and SSF systems. While the production cost in the crude fermentation by SmF was about $ 20/kg, by SSF it was only $ 0.2/kg if in situfermentation was used. The enzyme in SSF crude product was concentrated; thus it could be used directly in such agro-biotechnological applications as silage or feed additive, ligno-cellulosic hydrolysis, and natural fibre (e.g. jute) processing. A number of reports have appeared on microbial cellulase production in recent years. Nigam and Singh have reviewed processing of agricultural wastes in SSF systems for cellulolytic enzyme production. They argued that with the appropriate technology, improved bioreactor design, and operation controls; SSF may become a competitive method for the production of cellulases. They also enumerated advantages of cellulase production together with the factors affecting the cellulase production in SSF systems.

In a recent study on the ligninolytic system of Cerrena unicolor 062 – a higher basidiomycete – upon supplementation of the medium with carbon sources and phenolic compounds in SSF system, it was observed that the growth of C. unicolor 062 could be regulated by the exogenous addition of these compounds. The efficiencies of the degradation of cellulose and lignin were dependent on the nature and concentration of the compounds added53. Sun et al. developed a novel fed-batch SSF process for cellulase production which could overcome the problems associated with high initial nutrients concentration while retaining advantages from the high total effective salt concentration.

There are several reports describing co-culturing of two cultures for enhanced enzyme production. Gupte and Madamwa cultivated two strains of Aspergillus ellipticus and A. fumigatus and reported improved hydrolytic and b-glucosidase activities compared to when they were used separately using SSF system, improved enzyme titres were achieved by Kanotra and Mathur when a mutant of Trichoderma reesei was co-cultured with a strain of Pleurotus sajor-caju with wheat straw as the substrate. However, the media constituents too play an important role in mixed culturing. Gutierrez-Correa and Tengerdy reported that single culture of T. reesei and Aspergillus phoenicus, when supplemented with inorganic nitrogen source, produced similar xylanase levels as mixed cultures. However, when the fermentation medium was supplemented with soy meal, 35–45% more xylanase (than the single culture) was produced by these cultures.

In a significant finding, Smits et al. reported that glucosamine level of the fungi in liquid culture could not be used to estimate the biomass contents in SSF. They studied the SSF of wheat bran by T. reesei and reported that using glucosamine, correlation between the fungal growth and respiration kinetics could only partly be described with the linear growth model of Pirt. A decline in $O_2$ consumption rate (OCR) and $CO_2$ evolution rate (CER) started the moment glucosamine was 50% of its maximum value. After the glucosamine level reached its maximum, OCR and CER still continued to decrease.

A pan bioreactor, requiring a small capital investment, was developed for SSF of wheat straw. High yields of complete cellulase system were obtained in comparison to those in the SmF. A complete cellulase system is defined as one in which the ratio of the b-glucosidase activity to filter paper activity in the enzyme solution is close to 1.0. The prototype pan bioreactor however required further improvements so that optimum quantity of the substrate could be fermented to obtain high yields of complete cellulase system per unit space.

Although xylanases produced by fungi, yeast and bacteria, filamentous fungi are preferred for commercial production as the levels of the enzyme produced by fungal cultures are higher than those obtained from yeast or bacteria. In many microorganisms, xylanase activity has generally been found in association with cellulases, b-glucosidase or other enzymes, although there are many reports that have described in SSF systems, production of cellulase-free and other enzymes-free xylanase. Haltrich et al. reviewed the different factors that influence xylanase production by fungi. In view of the considerable commercial importance of enzymes, it was emphasized that efforts should be directed towards enhanced enzyme production with reduced associated costs.

Archana and Satyanarayana described a SSF process for the production of thermostable xylanase by thermophilic Bacillus licheniformis. Enzyme production was 22-fold higher in SSF system than in SmF system. Cai et al. Also reported production of a thermostable xylanase in SSF system. Enzyme produced in SSF system was more thermostable than in SmF system. Dunlop et al. described a bacterium, isolated from wood compost, producing xylanase that was active at 80°C. Jain too described a SSF process for the production of xylanase by thermostableMelanocarpus albomyces.

Alam et al. using SSF process, isolated a thermostable cellulase-free xylanase produced by T. lanuginosa. Addition of 0.7% xylan induced enzyme production to an extent of 28%. The enzyme was stable at 70°C. A thermostable xylanase preparation from Humicola sp. showed the temperature optima at 75°C. Srivastava reported a xylanase from Thermomonosporasp., which was stable at 80°C. Tuohy and Coughlan compared thermostable xylanase production on various substrates by a strain of Talaromyces emersonii in liquid culture and SSF systems. The latter showed higher enzyme activity compared to former, but liquid culture resulted in greater yields (U/g substrate).

Several authors have compared the performance of various microbial strains, grown on different substrates (individual or in combination) and reported varying results. Wiacek-Zychlinska et al. Compared xylanase production by C. globosum and A. niger on four different substrates. Although activities obtained by A. niger were higher than those from the other microbial cultures, but high-spore production by the A. niger strain could result in problems for a pilot plant or large-scale process.

In order to achieve improved enzymes titre, it is generally a common practice to pre-treat cellulosic or ligno-cellulosic substrates before using them in SSF systems. Pre-treatment may be by physical processes or chemical processes. Pre-treatment of palm oil mill waste, however, did not affect xylanase production.

b-xylosidase is another important enzyme used in textile industry. A b-xylosidase (EC-3.2.1.37) was produced by A. awamori K4 in SSF system on wheat bran, which was used for transxylosylation reactions. There are other reports as well describing the production of b-xylosidase in SSF systems.

### 6.14.5 Ligninases

Lignin is a three-dimensional phenylpropanoid polymer which is considerably resistant to microbial degradation in comparison to polysaccharides and other naturally occurring biopolymers. Biological delignification by SSF processes using microbial cultures producing ligninolytic enzymes – the ligninases – can have applications in delignification of ligno-cellulosic materials, which can be used as the feedstock for the production of biofuels or in paper industry or as animal feedstuff. These may also be used in pulp bleaching, paper mill wastewater detoxification, pollutant degradation, or conversion of lignin into valuable chemicals.

Lignin peroxidase (LiP, EC-1.11.1.7), manganese peroxidase (MnP, EC-1.11.1.13) and laccase (EC-1.10.3.2) are the most important lignin-modifying enzymes. LiP and MnP are heme-containing glycoproteins requiring hydrogen peroxide as an oxidant. LiP oxidizes nonphenolic lignin structures by abstracting one electron and generating cation radicals, which are then decomposed chemically. MnP oxidizes Mn(II) to Mn(III), which then oxidizes phenolic compounds to phenoxy radicals. This leads to the decomposition of the lignin substructure. Laccase, a copper containing oxidase, utilizes molecular oxygen as the oxidant and oxidizes phenolic components to phenoxy radicals.

Literature survey shows that a number of microorganisms produce ligninases96–112, but white-rot fungi generally show the most desirable qualities, in particular Pleurotus species and Phanerochaete chrysosporium are the most widely studied.

Wheat straw was used for cultivating several fungal strains to produce laccase, Li-peroxidase, and Mn-peroxidase. Several authors have used bagasse also Homolka et al.96 studied laccase production from three strains of Pleurotus sp. (obtained after protoplast regeneration of the control strain). While two strains showed significantly higher laccase activity, one strain showed lower activity. The rate of mineralization of 14C-lignin in SSF system by the latter and the control strain were almost the same, but it was higher than that of the other two strains. 14C-lignin in SSF of wheat straw was also used by Camarero et al. for studying Mn-mediated lignin degradation by four strains of Pleurotus sp., and comparing with by P. chrysosporium. At the end of the incubation period, strains of Pleurotus sp. acquired higher delignification values than P. chrysosporium. All the species of genus Pleurotus, studied so far, produce Mn-peroxidase, laccase, and aryl-alcohol-oxidase.

Dombrovskaya and Kostyshin studied the effects of different ionic nature surfactants on ligninolytic enzyme complexes of the white-rot fungi in SSF processes. The cationic surfactant, ethonium, enhanced the laccase and Mn-peroxidase activity by 1.8 fold and 1.6 fold, respectively for P. floridae. Kerem and Hadar studied the effects of Mn on the production of ligninolytic enzyme complexes of P. ostreatus in a chemically defined SSF system. Laccase, Mn-peroxidase, and catalase activities, and $H_2O_2$ production were all affected by Mn levels.

Laplante and Chahal compared ligninase production in SmF system and SSF system using a culture of P. chrysosporium ATCC 24725. Higher yields of ligninases, especially laccase and Mn-peroxidase, were obtained in SSF system. Kerem et al. compared the ligninolytic activity of a strain of P. chrsosporium BKM with P. ostreatus Florida f16. The former grew vigorously resulting in rapid, non-selective degradation of 55% of the organic components of the cotton stalks within 15 days. P. ostreatus grew more slowly with obvious selectivity for lignin degradation, resulting in the degradation of only 20% or the organic matter in 30 days.

### 6.14.6 Proteases

Proteolytic enzymes account for nearly 60% of the industrial market in the world. They find application in a number of biotechnological processes, viz. in food processing and pharmaceuticals, leather industry, detergent industry, etc. Recently, Mitra et al. reviewed production of proteolytic enzymes in SSF systems. From their viewpoint, proteases produced by SSF processes have greater economic feasibility.

In recent years, there have been increasing attempts to produce different types of proteases (acid, neutral, alkaline) through SSF route, using agro-industrial residues. It is interesting to note that although a number of substrates have been employed for cultivating different microorganisms, wheat bran has been the preferred choice in most of the studies. Malathi and Chakraborty evaluated a number of carbon sources (brans) for alkaline protease production and reported wheat bran to be the best for cultivation of A. flavus IMI 327634. Studies were carried out to compare alkaline protease production in SmF systems and SSF systems114. The total protease activity present in one-gram bran (SSF) was equivalent to 100-ml broth (SmF). A repeated batch mode SSF process was described for alkaline protease production in which polyurethane was used as the inert solid support. A thermostable alkaline protease was reported to be produced by a novel Pseudomonas sp. in SSF system. A process has been developed at CLRI, Chennai (India), for the commercial production of an alkaline protease (Clarizyme) which was produced by SSF of wheat bran using a strain of A. flavus.

A new strain of A. niger Tieghem 331221 produced large quantities of an extra-cellular acid protease when grown in SSF system using wheat bran as the sole substrate. Various C-sources inhibited protease synthesis, indicating the presence of catabolic repression of protease biosynthesis. The enzyme showed potential for usage as a bating agent. Ikasari and Mitchell used rice bran for acid protease synthesis by a strain of R. oligospora. They observed that although the enzyme showed optimum activity at pH 4, a leaching solution of pH 7 gave the optimum recovery of the enzyme from the fermented matter. They made stepwise changes in the gas environment and temperature during SSF process to mimic those changes which arose during SSF due to mass and heat transfer limitations. It was observed that a decrease of $O_2$ concentration from 21% to 0.5% did not alter protease production. Yaoxing et al. carried out SSF of wheat bran with a strain of A. niger QX 1066 for acid-resistant protease. High enzyme activities were obtained in a medium containing high carbon and low nitrogen content. Addition of a suitable phosphate in the medium further improved the enzyme titres. Villegas et al. studied the effects of $O_2$ and $CO_2$ partial pressure on acid protease production by a strain of A. niger ANH-15 in SSF of wheat barn. Results showed a direct relationship between pressure drop, production of $CO_2$, and temperature increase. Acid protease production increased when the gas had 4% $CO_2$ (v/v), and it was directly related with the fungus metabolic activity as represented by the total $CO_2$ evolved.

Germano et al. used a strain of P. citrinum for serine protease production using agro-industrial residues. The strain also exhibited lipase activity. Datta used aspen wood for the production of protease from the fungal strain of P. chrysosporium BKM-F-1767. Study of this enzyme's characteristics showed that this protease had properties of aspartate-type protease as well as of thiol-type protease.

### 6.14.7 Lipases

Fat splitting has been completely revolutionized by the introduction of lipases (EC-3.1.1.3) into the industrial arena. The conventional physico-chemical means of lipolysis have now been undershadowed by the biocatalysis using microbial lipases. Lipases have a wide array of industrial applications in the production and processing of detergents, oils, fats and dairy-products. In addition, they are also used in the preparation of therapeutic agents.

Until recently, SmF was in vogue for microbial lipase production. However, in recent years the shift has been towards the study and development of lipase production in SSF system1. Beuchat investigated SSF of peanut press-cake using Neurospora sitophila and Rhizopus oligosporus. Rivera-Munoz et al. compared SmF systems and SSF systems for lipase production using several filamentous fungi. Enzyme titres by SSF processes were higher and stable. Among the tested microbial strains, P. candidum, P. camembertii, and M. miehei proved the best for lipase production.

Benjamin and Pandey and Benjamin cultivated Candida rugosa on coconut oil cake for lipase production using SSF and SmF systems. Enzyme yields were higher in the former. Several carbon sources – individually and in combinations – were tested for their efficiency to produce lipases. Raw cake supported the growth and lipase synthesis by the yeast culture. However, supplementation with additional C- and N-sources increased enzyme titres. In contrast to this, however, Ohnishi et al. reported less lipase production from A. oryzae using SSF compared to SmF where high enzyme yields were obtained. Yet, in another comparative study on lipase production in SmF and SSF systems, Christen et al. observed a 5-fold increase in lipase productivity in SSF system.

Bhusan et al. reported lipase production in SSF system from an alkalophilic yeast strain belonging to Candida sp. Rice bran and wheat bran, oiled with different concentrations of rice bran oil were used as the substrate. Rice bran supplemented with oil gave higher lipase yields. Ortiz-Vazquez et al. and Granados-Baeza et al. used wheat bran for cultivating the strains of P. candidum. They designed an enzyme-recovery procedure and reported that 0.01 M NaCl was adequate to recover enzyme from the fermented matter.

### 6.14.8 Pectinases

Studies have been conducted on comparative production of pectinases in systems of SmF and SSF. When the fermentation medium was supplemented with different carbon sources, like glucose, sucrose and galacturonic acid, polygalacturoanase (PG, EC-3.2.1.15) production by A. niger $CH_4$ increased in SSF system but decreased in SmF system. Overall productivity by SSF was 18.8 and 4.9-fold higher for endo-PG and exo-PG, respectively, than those obtained by SmF. Minjares-Carranco et al. made physiological comparisons between pectinase-producing mutants of A. niger C28B25, adapted either to SmF or SSF. A. niger produced isozymes with difference in PG properties depending on the culture technique and strain used. The results also suggested that pleiotropic mutations of different kinds simultaneously affect the sporulation and enzymological patterns of each class of mutants.

Media acidity plays a significant role on pectinases' production by SSF processes. Cavalitto et al. and Hours et al. studied growth and pectinase production by A. foetidus and A. awamori, respectively in SSF systems at different media acidities. Both used wheat bran as the substrate. Results showed that higher the HCl concentration used, higher was the total pectolytic activity achieved. The low pH of the culture condition maintained asepsis during fermentation.

Apart from wheat bran, several other substrates have also been used for pectinase production in SSF system. These include coffee pulp citrus waste and apple pomace. Huerta et al. used bagasse as the inert substrate to produce PG in a 130 litres-packed bed fermenter byA. niger $CH_4$ (they referred it as 'absorbed substrate fermentation'). They claimed that the process was an efficient one for PG production as well as an interesting model since the culture medium, water, nutrients and specific inducers could be varied depending on the concentrations required. Acuna-Arguelles et al. studied effect of water activity (aw) on exo-pectinase production by A. niger $CH_4$ in SSF system. Sugar cane bagasse was used as the (inert) substrate and ethylene glycol was used as the water activity depressor. Results showed that although PG production decreased at low aw values, the activity was present even at as low as 0.90 aw values. The specific activity increased up to 4.5 fold by reducing the aw from 0.98 to 0.90.

### 6.14.9 Galactosidases

There has been considerable interest to produce a-galactosidase (EC-3.2.1.22) and b-galactosidase (EC-3.2.1.23) in SSF processes. Both these enzymes have applications in the pharmaceutical and food industries.

Cruz and Park reported production of a-galactosidase in SSF system and its application in the hydrolysis of galactooligosaccharides in soybean milk. Addition of soybean carbohydrate in the fermenting medium, using A. oryzae, was shown to induce enzyme production. Annunzaiato et al. carried out SSF of wheat bran for a-galactosidase production using a strain of A. oryzae QM 6737 with the aim of improving enzyme yields and lowering production costs. Enzyme yield increased 3 fold when soy flour or soybeans were used as the substrate, but no enzyme was produced using rice. Somiari and Balogh used a strain of A. niger for a -galactosidase production on wheat bran or rice bran. Srinivas et al. described the use of Plackett–Burman design for rapid screening of several nitrogen sources, growth/product promoters, minerals and enzyme inducers for the production of a-galactosidase by A. niger MRSS 234 in SSF.

In 1990, Wakamoto Pharma patented (two patents) the production of b-galactosidase in SSF systems. Strains of Aspergillus sp. and Penicillium sp. were used. Details have been provided in these patents by giving an example of the cultivation conditions and yields using a strain of A. oryzae. Enzyme preparation from A. fonsecaeus, which was cultivated on wheat bran, showed superior qualities than the other commercial preparation using a strain of A. oryzae and the enzyme was more suitable for biotechnological applications. Gonzalez and Monsan also used a strain of A. fonsecaeus for b -galactosidase production by SSF of wheat bran.

A thermostable b-galactosidase was reported from a thermophilic Rhizomucor sp. Enzyme activities by SSF were 9-fold more than by SmF processes. Strains of Kluyveromyces sp. have also been employed for b-galactosidase synthesis in SSF systems. Becherra and Sisocultivated K. lactis NRRL T-1140 on corn grits and wheat bran in SmF and SSF systems. They observed that change from liquid to solid state culturing did not promote b-galactosidase secretion by the yeast strain, though there were problems of drying of medium etc. in SSF. However, studies on production of b-galactosidase in SSF systems had already been published in 1995.

### 6.14.10 Glutaminases

L-glutaminase is considered a potent anti-leukamic drug and has found application as a flavour-enhancing agent in food industry. In a maiden report, Prabhu and Chandrasekaran reported L-glutaminase production by SSF using marine Vibrio costicola. Polystyrene was used as the inert substrate. They also evaluated several organic substrates for their ability to produce glutaminases by SSF using the same strain. Among the tested materials, wheat bran and rice bran were found superior in comparison to saw dust, coconut oil cake, and groundnut cake. However, use of polystyrene as the substrate offered several advantages over organic substrtes. For example, leachate from polystyrene-SSF system was not only less viscous but also showed high specific activity of the enzyme.

### 6.14.11 Amylases

The amylase family of enzymes has been well characterized through the study of various microorganisms. Presence of two major classes of starch-degrading enzymes have been identified in the microorganisms, viz. a-amylase (endo-1,4-a -D-glucan glucohydrolase, EC-3.2.1.1) which randomly cleaves the 1,4-a -D-glucosidic linkages between the adjacent glucose units in linear amylose chain, and glucoamylase (synonym amyloglucosidase – also referred to as glucogenic enzyme, starch glucogenase, gamma amylase; exo-1,4-a -D-glucan glucanohydrolase, EC-3.2.1.3) which hydrolyses single glucose units from the nonreducing ends of amylose and amylopectin in a stepwise manner. Unlike a-amylase, most glucoamylases are also able to hydrolyse the 1,6-a-linkages at the branching points of amylopectin, although at a slower rate than 1,4-linkages.

Amylases and glucoamylases are produced by various microorganisms, including bacteria; fungi and yeast, but a single strain can produce both these enzymes as well. These enzymes have found applications in processed-food industry, fermentation technology, textile and paper industries, etc. Selvakumar et al. reviewed microbial synthesis of starch-saccharifying enzymes in solid cultures.

SSF has been employed to produce amylases. In a recent study, Ray et al. compared the production of b -amylase (EC-3.2.1.2) from starch waste by a hyper-amylolytic strain of Bacillus megaterium B6 mutant UN12 by SmF and SSF processes. The starchy wastes used as substrates were from arrowroot, arum, maize, potato, pulse, rice, rice husk, tamarind, kernel, cassava, water chestnut, wheat and wheat bran. Arum and wheat bran gave the highest yields.

Comparative studies on a-amylase production using different substrates have been studied as well1. A new source of a -amylase was identified in Pycnoporus sanguineus. Cultivation of it in SSF system resulted in 4-fold higher enzyme production than in SmF system. Krishna and Chandrasekaran cultivated Aeromonas caviae (CBTK 185) on banana waste. The results indicated excellent scope for utilizing this strain and banana waste for commercial production of a -amylase by SSF. Sudo et al. compared acid-stable a -amylase production in SmF and SSF systems to ascertain as to why A. kawachii IFO 4308 produced larger amounts of acid-stable a -amylase in SSF system than in SmF system. Some of the attributes of SSF system were reported as the major reasons for higher enzyme production by SSF. A comparative study on SmF and SSF of inert substrate using a strain of A. oryzae CBS 125-59 also showed superiority of SSF system.

Lonsane and Ramesh reviewed the production of bacterial thermostable a -amylases by SSF, which they referred to as the potential tool for achieving economy in enzyme production and starch hydrolysis. Various methods to reduce the cost of production were discussed, taking into consideration enzyme production by B. amyloliquefaciens and B. licheniformis.

Numerous other microorganisms like Saccharomycopsis capsularia, B. coagulans1 Bacillussp. HOP-40186, and B. megatarium 16 M have also been used for a -amylase production by SSF using agro-industrial residues.

Recovery of the enzymes from the fermented matter is an important factor that affects the cost-effectiveness of the overall process. In a significant finding, Padmanabhan et al. reported that the recovery of a -amylase from the solid fermented matter depended on the temperature of extraction. When enzyme was extracted and recovered at 50°C, the quantum of recovery was 2.2 fold higher than at 30°C. A further increase of about 19% in leaching efficiency was observed when contact time was extended from 60 to 120 min.

The other important enzyme of the amylase family is glucoamylase (GA). Traditionally, glucoamylase has been produced by SmF and one-way process in solution has been well developed. In recent years, however, the SSF processes have been increasingly applied for the production of this enzyme.

A strain of A. niger was used for the production of glucoamylase in solid cultures. The study included screening of a number of agro-industrial residues including wheat bran, rice bran, rice husk, gram flour, wheat flour, corn flour, tea waste, copra waste, etc., individually and in various combinations. Apart from the substrate's particle size, which showed profound impact on fungal growth and activity, substrate-moisture content and water activity also significantly influenced the enzyme's yield. Different types of bioreactors were used to evaluate their performances. These included flasks, aluminium trays, and glass columns (vertical and horizontal). Enzyme production in trays occurred optimally in 36 h in comparison to typically required 96 h in flasks. In a significant study on the effect of yeast extract on glucoamylase synthesis by A. niger NCIM 1248 in SSF system, it was observed that supplementation with 0.5% yeast extract resulted in about 20% increase in enzyme yields. GA was purified 32.4 fold with the final specific activity of 49.25 U/mg protein. Four different forms (GA-I, GA-I', GA-II, and GA-II'), having different characteristics were reported. This was the first report on the four forms of GA produced by A. niger by SSF.

There are reports describing a comparative profile of glucoamylase production in SmF and SSF systems. Interestingly, contrary to the general findings, Fujio and Morita reported a 4.6-fold lower glucoamylase yield by Rhizopus sp. A-11 in a conventional SSF process using wheat bran medium than by SmF which used metal-ion supplemented medium. Solid and liquid cultures yielded 150 and 189 mg of protein, respectively. Hata et al. compared the two glucoamylases produced in SmF and SSF systems using A. oryzae. Enzyme produced by SSF could digest raw starch but that by SmF could not. GA obtained by the two systems exhibited different characteristics. Tani et al. too compared characteristics of GA produced by either SmF and SSF processes. Solid culture was more efficient than liquid culture for GA production.

Rajgopalan et al. used a bacterial strain of B. coagulans for modelling of substrate-particle degradation in SSF system of GA. Enzyme diffusion was found to be a critical factor in degradation of the substrate particle. Mitchell et al. studied an empirical model of growth of R. oligosporus in SSF system. An equation was developed to describe glucoamylase activity on the substrate, which was then used to predict the growth. Apart from an early discrepancy, the growth rate correlated reasonably with the GA activity. Elegado and Fujio screened 39 Rhizopus isolates and 9 authentic Rhizopus strains (grown on wheat bran in a SSF system) for their soluble starch digestive GA (SSGA) and raw starch digestive GA (RSGA) activities. Results showed that these strains could be classified into four groups, based on their SSGA and RSGA production and ratio of SSGA to RSGA. Soccol et al. also screened 19 Rhizopus strains for their ability to grow on raw cassava. Only three strains grew significantly, and GA production was higher on raw cassava than on cooked cassava.

A patent was granted to Snow Brand Milk Prod in 1990 for a process for GA production on multi-stage culture medium. An effective method for GA production in SSF was also described by Kobayashi et al There are many other reports on GA production in SSF systems using different strains on various substrates.

Critical analysis of the literature shows that production of industrial enzymes by SSF offers several advantages. It has been well established that enzyme titres produced in SSF systems are many-fold more than in SmF systems. Although the reasons for this are not clear, this fact is kept in mind while developing novel bioreactors for enzyme production in SSF systems. It is hoped that enzyme production processes based on SSF systems will be the technologies of the future. Genetically improved strains, suitable for SSF processes, would play an important role in this.

## 6.15 FERMENTATION IN THE MEAT INDUSTRY

Fermentation plays a big role in the meat industry as it is used during the process of creating cured sausages such as salami. These sausages are made up of a batter containing lean and fat meat, spices, sodium chloride, sugars and starter cultures – a starter culture is a dehydrated set of microorganisms used as a catalyst in the culturing process. This batter is then stuffed in natural or synthetic casings and placed into fermentation rooms. In these rooms, parameters such as temperature and humidity are carefully controlled to allow for optimal fermentation conditions. The microbe activity of the flora that is naturally present in the meat influences the colour, taste and consistency of the final product after fermentation. Modern food-processing technology makes it possible to follow and modulate the fermentation process from start to finish. This way, companies can ensure the quality and consistency of each and every product.

There are many positive effects of meat fermentation, such as:

- The drop in pH level, especially after homolactic fermentation, which produces lactic acid and fumaric acid
- The suppression and control of harmful pathogens
- The effects on the organoleptic characteristic
- The thickening of the meat mass due to the formation of the "gel" from the proteins
- The formation and stabilization of the typical bright red colour
- The reduction of Free water (Aw)
- The creation of the specific fermented taste, which is the result of enzymatic reactions of proteolysis and lypolisis

Klever Equipped works with Travaglini to bring you the most advanced and high-tech fermentation rooms. Thanks to the combination of the installed cooling and heating elements, it is possible to remove a noticeable amount of water from the product, even right at the beginning of the fermenting process, at relatively low temperatures. This way, water activity is reduced, fermentation is controlled and negative effects are prevented. The risk of crust formation is averted because the equipment is driven by the moisture released from the product. Having the equipment alternate between working and resting, and being able to monitor specific psychrometric parameters (easily displayable through graphics), increases the quality of the fermenting process of the salami.

## 6.16 DOWNSTREAM PROCESS IN FERMENTATION

The recovery and purification of fermentation products is one of the most important aspects of industrial fermentation processes. The selection of suitable process of recovery and purification depends upon the nature of the end product, their concentration, the by-products present, the stability of the product and degree of purification.

However, the selected procedure must be quick, simple, reliable, accurate and must measure the desired compound accurately from among the different chemicals that may be present in the growth medium. It is generally considered that the fermentation and product recovery are integral parts of an overall process. Because of interaction between the two, neither of them be developed independently, is necessary as this must not result in problems and unnecessary expenditure.

**General Pattern of a Recovery Process:** The following flow chart shows the various stages that are generally involved in the recovery process of a biological product. The recovery of an extracellular product of a fermentation broth is taken as an example to describe the various stages involved in a recovery process (Fig. 6).

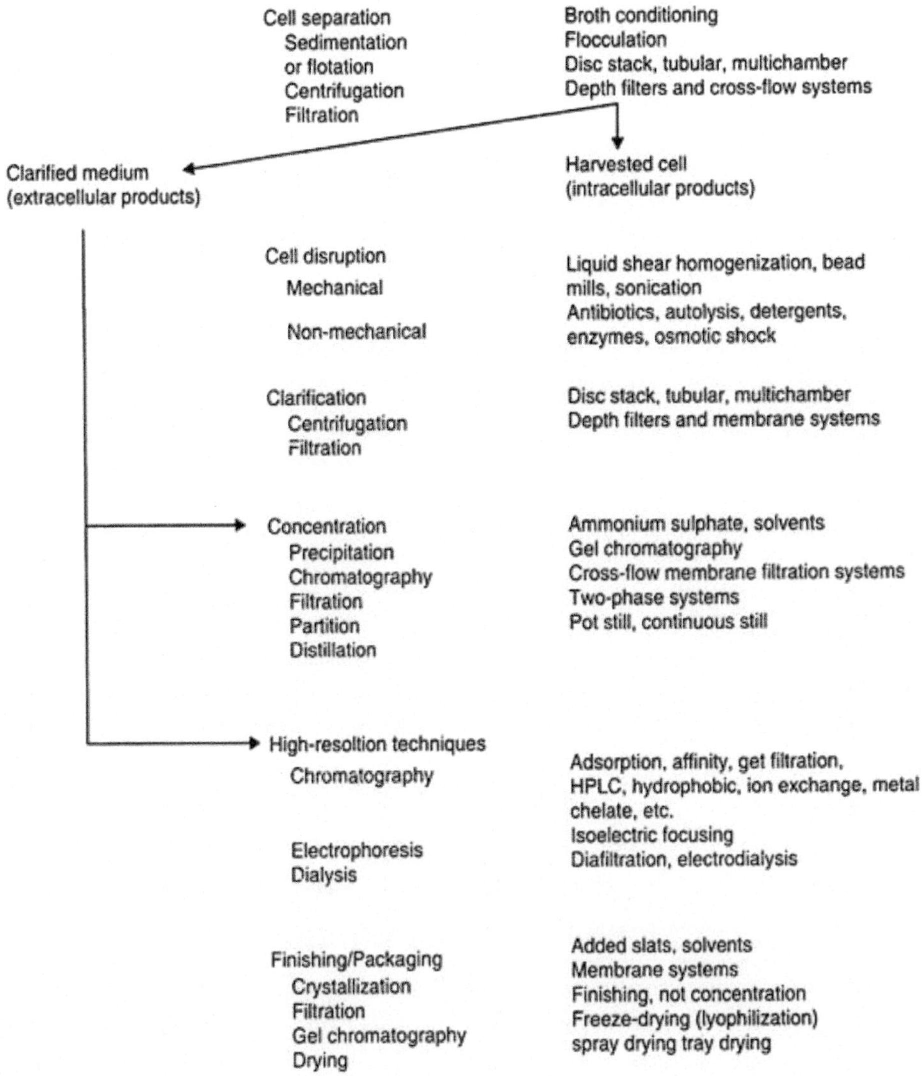


**Fig. 6. Typical unit processes used in downstream processing.**

The first stage for the recovery of an extracellular product is the removal of the large solid particles and microbial cells by centrifugation or filtration process. The broth is then fractionated or extracted into major fractions. Fractionation is carried out by employing either ultrafiltration or reverse osmosis or adsorption/ion exchange/gel filtration or affinity chromatography or liquid – liquid extraction or two-phase aqueous extraction based on chemical and physical properties of substances or precipitation.

The product-containing fraction is then purified by fractional precipitation. Further, more precise chromatographic techniques or crystallization process is employed to obtain a product, which is highly concentrated and devoid of any impurities.

**Various recovery processes can broadly be classified into the following categories:**

1. Physico-chemical processes.
2. Chromatographic partition processes.
3. Biological assays.

**1. Physico-Chemical Processes:** Though there are many physico-chemical recovery processes employed for the isolation of fermentation products, major of them are described here. However, it is to be remembered that the actual choice of recovery process largely depends upon chemical reaction or chemical analysis involved because the fermentation broth contains many compounds in addition to the desired one.

**(a) Turbidity Analysis:** This technique is employed for the determination of cell yield or biomass, which is the primary product of the fermentation process. Centrifugation and determination of packed cell volume is the simplest method by which estimation of cell yields of a microorganism can be made quickly. Fermentation broth with the cells is centrifuged in graduated centrifuge tubes and the volume of sediment cells are measured in centimeters.

The other method employed for measuring the cell yields is the turbidity analysis of a fermentation broth, containing little insoluble debris other than cells. The cells present in the growth medium are diluted to a turbidity range and the cell suspension is exposed to a range of visible light or a monochromatic light with wave-length (660 nm) in a colorimeter. Quantitative cell measurements are made as optical density by deflection of light that is caused by the microbial cells present in the diluted suspension.

For determining number of cells, trubidimetric measurements of cell numbers are usually standardized against some other procedures, such as plate count. A standard curve is then drawn with the help of values of optical density of plate count which may be derived from a series of dilutions of the cell suspension. Once the standard curve is prepared, it is easy to determine the cell number by converting reading of optical density.

**(b) Titration and Gravimetric Analysis:** This technique is generally employed for the recovery of organic acids and the choice of recovery method varies according to the nature of the acid. For example the amount of an organic acid like lactic acid produced in fermentation can easily be determined by adding bromothymol blue or other pH indicating dye to a sample fermentation broth, followed by titration with an alkali of known concentration.

If the acid formed in a fermentation broth is able to form an insoluble salt with an alkali it can be isolated by precipitation, which can be washed, dried and weighed. Volatile acids with low molecular weight can be isolated by direct distillation of fermentation broth and distillate is analyzed by titration.

Similarly acids of high molecular weight can be separated from fermentation broth by adsorption to a suitable anion exchange resin and subsequent elution.

**(c) Spectrophotometry Determinations:** Different types of spectrophotometers are used during spectrophotometric determinations. The difference in these analytical instruments lies in the kind of light absorbed. Visible light spectrophotometers absorb visible light in the range of 350 to 1000 nm. They measure the amount of light at a specific wavelength which is absorbed as light beam passes through a colored solution.

Ultraviolet spectrophotometers absorb light within the wavelength range of 200 to 380 nm, while in a fluorescent spectrophotometer the amount of light emitted by a fluorescent compound is measured.

Fermentation products which are colored can be measured directly by visible light spectrophotometers. A specific wavelength of visible light is allowed to be absorbed by a fermentation product and the absorption of light is measured as optical density. A standard curve is prepared with the help of the optical density of known concentration of a pure concerned chemical. By substituting the values of optical density of the fermentation product in the standard curve, the amount of fermentation product can be determined.

The colorless fermentation products are made coloured by treating with a colouring reagent. For example – amino acids are generally colorless chemicals but they develop purple colour when treated with ninhydrin under suitable conditions. The amount of fermentative production of amino acid is then determined by following the method described above.

Some fermentation products are basically colourless and do not develop colours even after chemical treatment. Determinations of such products are made with the help of ultraviolet spectrophotometers. By allowing them to absorb wavelength of ultraviolet light and measuring the optical density of the absorbed light.

Still some fermentation products like riboflavin fluoresce when exposed to ultraviolet light. The intensity of fluorescent light is measured by employing fluorescent spectrophotometers or fluorometer. By determining the optical density of the fluorescent light, the amount of fermentation product can be determined. The principles of analysis for ultraviolet and florescent spectrophotometers are similar to those for visible spectrophotometers.

**2. Chromatographic Partition Determinations:** The term chromatography was originally used in earlier times for a technique to separate coloured compounds. Although partition is the predominant cause of separation, other factors are also involved though in a minor fashion.

Employment of partition chromatography on paper or thin layer plates, not only made marked strides in the fermentation research and technology, but also allowed to detect and identify many types of new fermentation products even though they are in extremely smaller amounts. An unknown compound can easily be detected with the help of this technique by comparing the partition value with already known compounds.

The basic principle involved in the partition chromatography is the continuous partition of the solute or the sample between a stationary phase, such as paper or the silica gel thin layer plates, and a mobile phase consisting of a mixture of solvents, as these solvents move across the paper or silica gel layer.

Basically there are eight types of partition chromatography:

(i) Paper chromatography

(ii) Thin layer chromatography

(iii) Column Chromatography

(iv) Ion Exchange Chromatography

(v) Adsorption Chromatography

(vi) Affinity Chromatography

(vii) Gel Filtration

(viii) Gas Chromatography

**(i) Paper Chromatography:** If a cellulose paper is used as a stationary phase it is called as paper chromatography. Water soluble compounds are generally separated by employing paper chromatography.

In this technique the fermentation product to be separated is applied as series of spots about ¾ apart along a line, about few centimeters away from one end of a rectangular filter paper. Sometimes sample is also applied as a single spot on the paper. The paper is kept dipped in a mixture of solvent system consisting of a hydrophobic and a hydrophilic solvent. The solvent mixture starts ascending the paper and thereby wetting it by capillary action.

But the cellulose, which is the component of a paper, can absorb only the hydrophilic solvent and not the other. Thus there exists a phase separation at the micro-level. The compounds present in the fermentation broth get gradually separated when the solvent front reaches the spot, which contains the fermentation products to be separated. As the solvent front moves further the fermentation products get separated as consolidated spots on the filter paper.

For carrying out the solvent migration on the filter paper, (also called as chromatogram,) it is placed in a closed tank so that its atmosphere becomes saturated with the components of solvent system and helps in the perfect separation of the product. The developed chromatogram is removed from the tank and allowed air dry to remove extra solvent. The separation is measured in terms of a unit called Rf (Resolution factor) which is described as relative to front.

Based on the method adopted, the processes of paper chromatography are classified into following categories:

**(a) Ascending Paper Chromatography:** In this method, the solvent system is placed as a shallow layer at the bottom of the tank and the filter paper strip is suspended in the solvent system in such a manner that the spotted sample does not dip in the solvent system, so that solvent ascends the paper (Fig. 7).

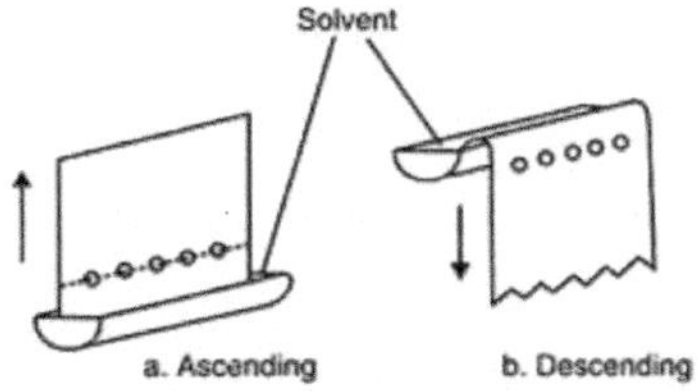


**Fig. 7. (a) Ascending and (b) Descending Chromatography**

**(b) Descending Paper Chromatography:** In this method, a trough containing solvent mixture is placed near the top of the tank and spotted chromatographic paper is placed in this trough so that the solvent migrates down the paper (Fig. 7).

**(c) Horizontal or Circular Chromatography:** In this method, a circular filter paper is chosen and a hole is made at its center. A wick placed in the central hole elevates the solvent to the chromatograph from a reservoir present beneath the sheet (Fig. 8) so that solvent migrates all round from the centre. This is also called as disc chromatography.

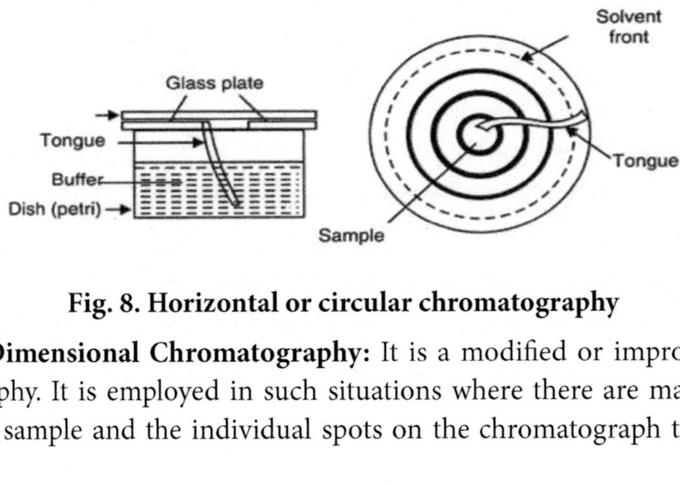


**Fig. 8. Horizontal or circular chromatography**

**(d) Two-Dimensional Chromatography:** It is a modified or improved method of chromato¬graphy. It is employed in such situations where there are many compounds present in the sample and the individual spots on the chromatograph tend to coalesce (Fig. 9).

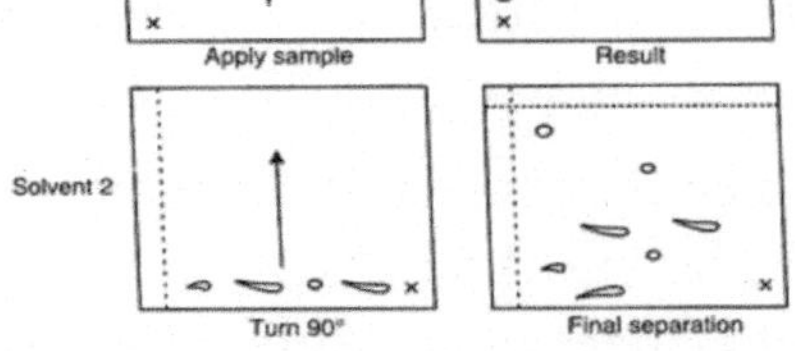


**Fig. 9. Two Dimensional paper Chromatography**

A single spot of the sample is applied at one corner of the rectangular filter paper and solvent migration is allowed across the paper. This moves out individual compounds along a line from the origin in the direction of solvent migration. Solvents are then evaporated by drying filter paper, it turned 90 degrees and placed in a second different solvent system.

Solvent migration with the second solvent system moves individual compound across the filter paper at an angle of 90° from the original direction of travel. This two dimensional chromatography allows the use of two different, even contrasting solvent system on a single chromatographic paper and separates the compounds more clearly.

However, it has certain major disadvantages are:

1. Only a single spot can be applied on the filter paper.
2. It is difficult to interpret the Rf values, as the values are associated with two different directions of solvent migration. But, it is still useful in the separation of compounds with small differences in their Rf values.

**(ii) Thin Layer Chromatography:** Thin Layer Chromatography (TLC) is similar in most respects to paper chromatography except the usage of a thin glass plate in place of filter paper. A thin glass plate is coated with a thin layer of silica gel or aluminum oxide, which acts as a stationary phase. Calcium sulfate or starch is generally incorporated into the stationary phase compound, which helps in binding the latter to the glass plate surface.

The layer is allowed to dry at this stage, it is called as chromatogram. Fermentation broth sample is applied as a series of small spots approximately ¾ inch apart along a line about one inch from one end of the glass plate. The spots are allowed to dry at room temperature or heated with a stream of warm air or an infrared lamp.

After the spots have dried up glass plate is kept in a tank, saturated with solvent mixture, in such a way that spot do not dip in solvent mixture. The solvent mixture is then allowed to migrate through the chromatogram. The individual compounds present in fermentation broth get separated based on their molecular weight as consolidated spots on the chromatogram, when solvent mixture migrates through it. For precise analytical determinations the temperature at which solvent migration occurs must be carefully controlled. When solvent migration is done TLC plates are allowed to air dry.

The compounds thus separated on the chromatogram can be detected with the help of suitable spray reagent or made to absorb ultraviolet light. Amino acids can be detected by spraying the chromatogram with ninhydrin to produce purple coloured spots. Radioactive compounds can be located by exposing the chromatogram to x-rays.

The separated compounds on the chromatogram can be identified by calculating the Rf values and subsequently comparing the values with the known compounds. Biologically active compounds such as antibiotics can be detected by placing the chromatogram for short period on the surface of an inoculated agar plate.

**(iii) Column Chromatography:** In this method, a vertical tube made up of glass or polyacrylate plastic is provided with an inlet and an outlet. The inlet is connected to reservoir containing a solvent to have a continuous flow of the solvent, while the outlet is fitted with rubber tubing to collect the eluting compounds. The column is fitted in the upright position and its bottom is sealed with a glass wool, which supports the stationary phase, which is gradually packed up to ¾ of the column.

The stationary phase is made of either gel or adsorbent or resin. The stationary phase is prepared in the form of a thick suspension and is called as slurry. After the slurry settles well in the column, the fermentation broth is applied at its top end to a height of 5-10 cm. The solvent mixture, which acts as a mobile phase, is made to flow through the column from its reservoir (Fig. 10).

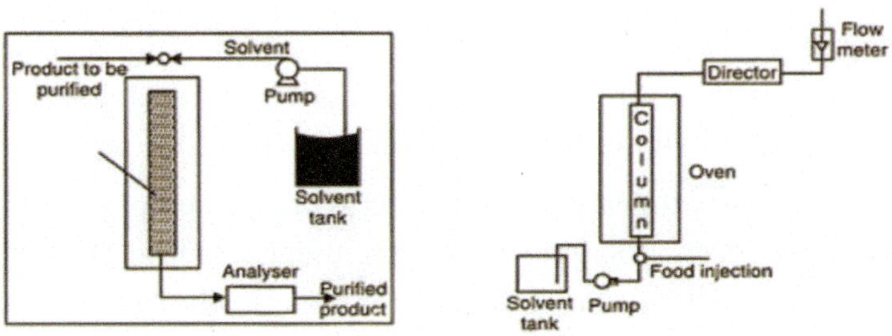


**Fig. 10. Basic Operation of Column Chromatography**

As the solvent mixture moves through the column the products present in the fermentation broth get separated as the products emerge from the column outlet they are analyzed, identified and calculations with regard to their quantity is also made by employing one of the several appropriate analytical methods available like fluorescence detectors, voltameters, refractometric detectors, conductivity detectors etc. This technique is used for the separation of proteins and related compounds.

**(iv) Ion Exchange Chromatography:** Macromolecules with ionic groups can be separated by ion exchange chromatography. The macromolecules are made to adsorb a carrier and are eluted by a solvent of defined strength. The separation mainly depends upon the nature of the macromolecule and the ionic strength of the eluting solvent.

This technique is generally used for the purification of antibiotics from fermentation broth, purification of proteins. The ionic material like Dowex, HCR and OCR, Amberlite, IR and IRC are widely used for cation exchange. Anion exchangers used are Dowex, SAR and MSA, Amberlite, IRA and Lewatit M.

**(v) Adsorption Chromatography:** In adsorption chromatography separation of the contents of biological material is achieved due to hydrophilic and hydrophobic reaction between them. Elution and fractionations are accomplished by means of solutions of higher or lower polarity or strength. Silicates, alumina, activated carbon, cross-linked dextrans and hydroxylapatite are generally employed as adsorption material.

**(vi) Affinity Chromatography:** It is a specific method for purification of biological material. The desired biological material binds specifically and reversibly to a ligand, which has been fixed to an inert carrier. For example – the Nucleotide Adenine Dinucleotide (NAD) can be purified by allowing it to bind to a carrier containing a dehydrogenase enzyme. The antibiotic bacitracin can be used as a ligand and to isolate aspartine, serine, cystine and metalloproteases from a crude mixture.

**(vii) Gel Filtration:** In this technique, a gel is used as filtering agent and molecules of different size can be separated by allowing them to pass through the gel with different pore size. As the large molecules cannot penetrate the gel and pass directly through the column, finally elute out along with the mobile front, whereas small molecules penetrate deep into the gel and hence their mobility is inhibited from the solvent front.

Standard curves can be developed which relate molecular weight to the elution position. This technique is mainly used at the industrial scale to eliminate salts and separate low molecular weight impurities. It is also used to purify proteins like insulin or interferon, at a smaller scale industrial production. The Biogel types P and A from BioRad and the Sephadex series S, G and LH from Pharmacia are the gels generally employed in gel filtration technique.

**(viii) Gas Chromatography:** In this technique fermentation products are first volatilized from the fermentation broth and then separated. The product to be separated from the fermentation broth is injected into the gas chromatograph and is converted into its gaseous state by heating.

The gas thus formed is pushed by a stream of inert gas through a partition column, and compounds get separated while passing through the column and are detected as they come out of the column by employing proper detecting system. Quantitative assay is obtained by comparison of the areas under peaks on plots of the data with corresponding areas for various quantities of reference compounds.

**3. Biological Assays:** Determining the effect of a fermentation product or an enzyme on the growth and metabolic activities of test organisms is called biological assay or bioassay. Therefore, it is an indirect method of assessing the quality of a fermentation product. Biological assays are, therefore, difficult to perform, less reproducible and usually do not give accurate results.

The microorganisms employed in these assays are called as test organisms. Strains of microorganisms occurring in nature or artificially mutated ones may be employed as test organism. Various categories of microorganisms are used as test organisms. For example, bacteria for assaying amino acids, antibiotics and vitamins; fungi for assaying vitamins and trace elements; yeasts for assaying vitamins and antibiotics.

A microorganism that is employed in a biological assay should possess the following characters:

1. It should genetically be stable and should not exhibit any genetic change in response to the test chemical.
2. It should respond in a graded manner to the test compound and not to other compounds.
3. It should not be a pathogen and should grow on simple medium.
4. It should preferably be aerobic or facultative aerobic but not an anaerobic.
5. It should grow well at a pH that does not affect the stability or toxicity of the fermentation product under assay.

**Types of Biological Assays:** Biological assays can be categorized into four general groups.

They are:

(i) Diffusion assays,

(ii) Turbidimetric assays,

(iii) Metabolic response assays and

(iv) Enzymatic assays.

**(i) Diffusion Assay:** In this assay, the fermentation product to be assayed is allowed to diffuse through medium in a radial fashion from a pad or cup and the adjacent growth of the test organism is either retarded or stimulated. The diameter of this area reflects the concentration of the compound being assayed and the amount of the product is determined by measuring and comparing the area with the various known concentrations of reference compounds.

**(ii) Turbidimetric Assays:** In this method, the effect of the fermentation product under test in a liquid culture is measured as an increased or decreased turbidity associated with the growth rate or total growth of the microorganism.

A suitable sterilized medium is taken in series of sterilized test tubes and graded amounts of the product to be assayed are added. The tubes are inoculated with an equal amount of vigorously growing young culture of the test organism and then incubated for a predetermined period of time at a constant temperature.

The length of incubation period depends on the type of growth measurements, that is, whether growth rate or total growth of test microorganism. For measuring the growth rate, the turbidity of the culture at some point during logarithmic growth is determined, while for measuring the total growth of organism the maximum stationary growth phase is determined.

The relative turbidity produced in the test tubes can be determined either visually or by employing a spectrophotometer. Readings are made as optical densities or absorbance. The optical density is plotted against the concentration of the standard to obtain a standard curve. The concentration of the unknown product is determined by comparing its optical density with the standard curve. (Fig. 11).

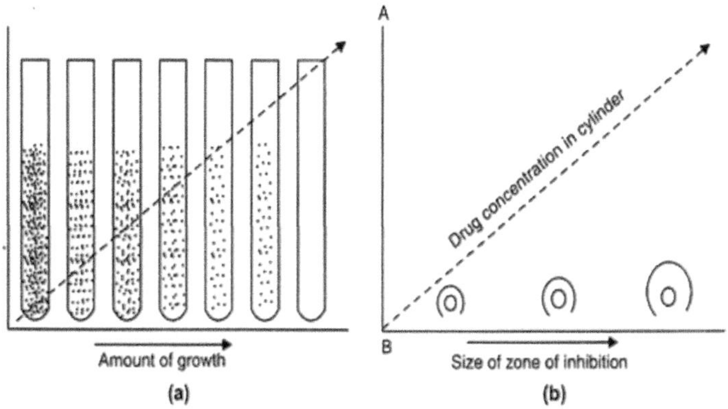


**Fig. 11. Antimicrobial Activity (a) Turbidimetric assay (Tube Dilution Technique) (b) Cylinder-plate technique**

**(iii) Metabolic Response Assays:** These assays are similar to turbidimetric assays except that, instead of measuring the effect of the fermentation product on the rate of growth or the total growth of the test organism, the effect on some metabolic reaction that test organisms carries out during growth are measured. A few metabolic reactions used for assay of this type are acid production, carbon dioxide evolution, oxygen absorption and enzyme dehydrogenase activity.

**(iv) Enzymatic Assays:** Enzymatic assays are more specific. They are able to detect quantitatively in extremely minute amounts of fermentable products and differentiate between biologically active and inactive forms of a compound. The required enzymes of these assays are derived from a commercial source or from a microbial culture or from other enzyme source.

The enzyme preparation is incubated with a sample of culture broth so as to cause some enzyme mediated changes in the fermentation product such as a partial decomposition with consequent formation of a measurable product.

For example – L-glutamic acid in a fermentation broth can be detected by adding cells of a pure culture of a strain of Escherichia coli, which contains the enzyme of glutamic acid decarboxylase. A little amount of toluene is also added to fermentation broth which facilities the release of enzyme from the cells of E. coli.

The test is carried out at an acidic pH 5.0. One molecule of $CO_2$ is generally liberated from each molecule of glutamic acid. The $CO_2$ liberated due to the degradation of L-glutamic acid by the enzyme glutamic acid decarboxylase collects in the atmosphere of the test tube containing the fermentation broth as a gas. Its concentration is calculated with a manometer such as Warburg respirometer.

# 7

# BEER AND WINE FERMENTATION

## 7.1 INTRODUCTION

Fermented beverages can be divided into two groups, wines and beers, broadly defined. Wines are fermented from various fruit juices containing fermentable sugars. Beers come from starch-containing products, which undergo enzymatic splitting by diastase, malting, and mashing, before the fermentable sugars become available for the yeasts and bacteria. The enzymatic splitting of the starch can also be performed either by human saliva, containing amylases, or by molds. Narrowly defined, beer is barley beer and wine is grape wine.

## 7.2 ALCOHOL FERMENTATION

Alcohol fermentation is done by yeast and some kinds of bacteria. These microorganisms convert sugars in ethyl alcohol and carbon dioxide. Alcoholic fermentation begins after glucose enters the cell. The glucose is broken down into pyruvic acid. This pyruvic acid is then converted to $CO_2$, ethanol, and energy for the cell. Humans have long taken advantage of this process in making bread, beer, and wine. In these three product the same microorganism is used: the common yeast or Saccharomyces Cerevisae

### 7.2.1 Wine Fermentation

Saccharomyces is responsible for the alcohol fermentation of wines. Grape juice contains naturally high levels of sugars. These sugars are converted into alcohol and carbon dioxide. Natural fermentation can yield wines with an alcohol up to 16 percent.

Alcohol is produced when yeast enzymes break up sugar into roughly equal parts of alcohol and carbon dioxide gas. Everyone may be familiar with bakers yeast but most yeast occur wild in nature and grow on plants and animals where they are dispersed through the air and water. Yeast spores are everywhere and if they get a chance they will gladly ferment your grape juice or malt. The problem with this is that there are thousands of strains of wild yeast and most of them are not suitable for fermenting alcohol.

Modern fermentation practices have all but eliminated the problems that plagued wine and beer makers for thousands of years. Our better understanding of yeast and bacteria have resulted in techniques that keep out unwanted microorganisms which can cause spoilage.

## 7.3 HOW TO MAKE PERFECT BEER AND WINE?

1. Use clean equipment in good condition. Dirt harbours bacteria and bacteria grows rapidly in wine and beer which can cause spoilage. Old discolored or scratched plastic equipment is impossible to clean and will cause problems.
2. Sanitize all equipment after cleaning using a specialized sanitizer supplied by home brew and wine shops.
3. Use a tight-fitting lid on all fermenting buckets and an air lock. Exposure to air can introduce wild yeast and bacteria. This is particularly important during the first 24 to 48 hours before your yeast has had a chance to start working.
4. Use only high-quality and fresh ingredients.
5. Use at least 10 grams of fresh yeast per five gallon batch of beer.
6. Follow a good set of instructions and don't listen to your next door neighbor "who has been brewing for years". I have seen too much beer and wine spoiled by well intentioned advice from neighbors and coworkers. If you have a question contact your wine or beer supply store. They deal with hundreds of queries and have the resources to answer your questions.

Fermented beverages have been produced and consumed all over the world and over a very long time span. Man discovered that sugar solutions of different origins, if left standing rather warm, will start fermenting spontaneously into an alcoholic beverage that also often contains lactic acid. The requisitemicroorganisms, Saccaromyces yeasts and Lactobacillus bacteria, are abundant almost everywhere and will do their duty, producing alcohol and lactic acid. A similar fermentation process of animal and vegetable foods is the lactic-acid fermentation that yields, for instance, sour herring in Sweden, and sauerkraut.

### 7.3.1 Beer from Cereals

Bouza is produced in Egypt and is probably the forerunner of beer in Ancient Egypt. It is prepared from malt of milo (a grain sorghum that resembles millet) and crushed baked loaves of bread. Residue from an earlier fermentation is used as a starter and both lactic-acid bacteria and Saccaromyces are involved in the fermentation process to get this sour alcoholic beverage. Talla from Ethiopia is a very similar to bouza and can be produced from barley and wheat. The bread loaves are heated to give a roasted character, and the pots are fueled by olive wood to give a smoky taste. Talla can also be spiced with hop leaves and stems, and spices.

**Wheat Beer:** Wheat beer—and rye and oat beer as well—are frequently made from mixtures of malt and the crushed grains of these cereals with barley. Often the beer is bottled with the yeasts for continued fermentation in the bottle. Wheat beer—which is top-fermented and thus, technically, an ale—is particularly popular in Bavaria and in Belgium and northern Germany, in varieties such as lambic, Gueuze, Wit beer, Trappist beer, and Berliner Weisse.

**Rye Beer:** Rye beers such as kalja and sahti in Finland, similar beverages in the Baltic area, and kvas ("kvass" in English) in Russia are produced in northern and eastern Europe, where cultivation of rye is widespread. Kvas has been the basic beverage for the Russian people for centuries. Its importance is indicated by many proverbs such as "Eat cabbage soup with meat, but if you don't have it, eat bread and kvas." The use of kvas is documented from about the year 800. Traditionally it is spontaneously fermented. In a broad sense, kvas is any sour and alcoholic fluid made from honey, bread, cereals, birch sap, fruits, beets, orcabbage. Red-beet kvas should always be used to produce borshch ("borscht," in English—red-beet soup). In a narrow sense, kvas (that is, bread-kvas) is produced from rye malt with or without other cereals or sour-fermented rye bread as adjuncts, and the beverage is spiced with peppermint. Most of the consumedkvas is of low alcohol content, only some few percent.

**Oat Beer:** Oat beer is today mostly used as an adjunct in certain stouts called oat stouts. A stout is adark ale, made of roasted malt, which occurs in several varieties such as sweet and dry (Guiness). Oats have also been used in traditional Norwegian beer production.

**Rice Beer:** Another type of cereal beers is made from rice, and these have been produced in all the rice-growing areas of the world. The best-known example is sake from Japan, which has a documented history of more than two thousand years. The principal difference from the malting of barley lies in the koji-process of rice. Koji is a culture of Aspergillus oryzae, which grows on steamed rice, and saccharifies the rice starch (that is, converts it to sugar) and decomposes the rice proteins. The sugar produced in the sake mash is later fermented by the already-present Saccaromyces cerevisiae. The mash is acidified either by adding lactic acid to it or by facilitating the growth of lactic bacteria to form the seed mash, moto. Sake is a clear, pale-yellow liquid with an alcoholic content of about 15 percent and a characteristic estery (artificially fruity) aroma. It is slightly sweet and slightly acidic and has a high amino-acid content compared to wine and beer. Other similar beverages from rice are known from other eastern and southern Asian countries such as China, Thailand, India, Malaysia, and the Philippines.

Another similar fermentation process is performed by the inoculum ragi, present in Southeast Asia. Ragi contains the mold Amylomyces rouxii, the yeast Endomycopsis burtonii, and sometimes Hansenula yeasts; it produces a pleasant alcoholic and acid beverage from rice or cassava, called tape ketan in Malaysia.

Sorghum and millet beer. Sorghum beers, known as kafir beers in Africa, are made from malt from sorghum or from the related grain, millet. Often-used adjuncts are maize (corn), malted or unmalted sorghum or millet, and malt amylase. Lactic acid is used as a flavoring and preserving agent, and the alcoholic fermentation is performed by Saccaromyces. Nowadays, these traditional African beers are not only produced in tribal areas; they are also available in home-brewed urban and industrially produced versions. They are opaque, rather thick pinkish-brown liquids with an estery or fruity odor, and a sweet and sour taste.

Maize beer. Tesquino and zendecho from Mexico and Latin America are made from malted maize and spontaneously fermented by Saccaromyces cerevisiae. These beverages might be as ancient as the oldest beers of the Old World, dating back about eight thousand years. Another beer is chicha, made by Andean and Central American Indians. The starchy material is chewed into dough, which is dried and later placed into warm water where the amylase action is finished. Then a starter is added, and the lactic-acid and alcoholic fermentation begins. Today much of the chicha is made using a maize malt rather than saliva.

### 7.3.2 Beer from Starch Products

All kinds of starchy items, such as manioc, potatoes, beets, and various roots, are included in this group. To get fermentable sugars, the starch has to be split either by diastase in malt, by saliva, or by molds, as with rice. In South and Central America, almost all of the traditional beers were originally produced by chewing either the cereal maize or other starchy vegetables. One of the most popular has been the manioc, both the sweet and the bitter; sweet potatoes, mangabeira, cashew, Jaboticaba (Myrciaria cauliflora), pineapples, bananas, and algarroba pods have also been used. In the tropical forest tribes, the favorite manioc beer was prepared as follows:

The roots, cut into think slices, were first boiled, then squeezed and partly chewed by young girls. The mass, impregnated with saliva, was mixed with water and heated again over the fire. The liquid was afterward poured into huge jars, half buried in the ground, covered with leaves, and left two to three days to ferment. A fire was built around the jars to warm the beverage before serving it. Each extended family manufactured its own liquor. When a bout was organized, drinkers went successively to each hut, exhausting the available supply. The women served the liquors in huge calabashes.

An earlier popular American low-alcoholic beverage is root beer, which consists of an infusion of sarsaparilla, sassafras, spruce, wild cherry, spikenard, wintergreen, and ginger, with sugar and yeast. Today, it is a soft drink containing some of these ingredients at its best; otherwise, it is artificially spiced.

### 7.3.3 Wine from Fruit and Vegetable Juices

**Fruit wine:** Fruit wines are produced with almost the same technique as grape wines. Specifically-named fruit wines such as cider (from apples) and perry (from pears) are produced, as well as wines from other fruits. The technical difference between these two groups is the alcohol content—5 to 7 percent in the first group and up to 18 percent in the second group—which depends on sugar addition.

**Tree-sap wine:** Saps from various trees have been used to produce alcoholic beverages; examples include maple sap (from Acer saccharum, the sugar maple) in North America, and birch sap (from Betula pubescens, the downy birch) in northern Europe. The manifestation of the symbiosis of yeast and bacteria cultures, which looks like a jellyfish, on the wounds of spring birches has been used in Europe as a folk medicine and is called "Volga-swamp."

**Sugarcane wine:** Wine made from the sugarcane, Saccharum officinarum, together with molasses, is distilled into rum.

**Cactus-plant wines:** The tall perennial plants of the genus Agave, which grow in Mexico and nearby areas, give a sweet, slightly bitter sap, called agua miel (literally, honey water), which is fermented into pulque, either spontaneously or through the use of an inoculum from a previous fermentation. The Aztecs were familiar with the product. Pulque is an important food beverage for the poor in the semiarid areas of Mexico. If pulque made from Agave tequilana is distilled, the resultant liquor is called tequila; if pulque made from another agave is distilled, it is called mescal. Before the contacts with the European settlers, only a few American Indian tribes north of Mexico made alcoholic beverages. They were the Akimel O'odham (Pima), Tohono O'odham (Papago), and the River Yuman peoples in southern Arizona and northern Mexico, and they produced wine from the saguaro cactus as well as from the agave and the mesquite. In the East, the Cherokees made wine from persimmons. During an important ceremony in July, the Papagos and the Pimas drank enormous quantities of the wine to induce rainfall in their desert areas.

**Sugar-palm wine:** Some examples of wines made from palm sap are surra from Borassus flabillifer, toddy and temba from the coconut palm Cocus nucifera (from which arrack is obtained by distillation), malovu from Elaeis guineensis, a kind of undistilled "rum" from Hyphaene coriaca and Hyphaene critina, Phonix reclinata, Raphia pedunculata, and Raohia vinifera.

### 7.3.4 "Wine" from Animal Sources

Mead. Honey is probably among the first foods gathered by Homo sapiens and its predecessors, and mead, the wine fermented from honey, may well be one of the oldest alcoholic beverages. Honey has also been much used as an adjunct to sweeten many kinds of beers over the centuries. Mead was the drink of the Nordic gods, whereas the people drank beer.

## 7.4 TYPES OF FERMENTATION THAT PRODUCE BEER & WINE

While beer and wine may taste completely different, they are produced by the same fermentation process. Alcoholic fermentation is a surprisingly simple process that converts sugar into alcohol and $CO_2$. Although the process of fermentation is the same for both beer and wine, the materials used and time needed can vary drastically.

**Alcoholic Fermentation:** Alcoholic fermentation converts sugars into carbon dioxide and ethyl alcohol. This is done by adding yeast to a substance rich in carbohydrates. The yeast actually eats the sugar, which reduces the overall sugar content of the finished alcoholic beverage. The carbon dioxide that is created during the fermentation process is released into the air while the alcohol remains in the beverage. The type of alcohol created in alcoholic fermentation is referred to as ethanol.

**Yeast:** There are many strains of yeast that are used in beer and wine. Each strain of yeast imparts a unique flavor to the alcoholic beverage. Each type of yeast also has specific alcohol tolerances. On average, beer yeast can only tolerate 5 percent ethanol while wine yeast can tolerate 12 percent. These numbers can fluctuate greatly depending on the type of yeast used. Yeast also relies on a certain temperature during the fermentation process. Once again, this varies based upon the strain of yeast used.

**Wine Ingredients:** When wine is created, yeast is added to grape juice and allowed to ferment. Wines typically have an alcohol content of about 12 percent, depending on the sweetness of the grape used and the style of wine that is desired. To make red wine, winemakers will crush the grapes and ferment both the juice and the skins. This makes the wine red in color and adds tannins, which gives the wine a dry taste. White wine is made by removing the grape skins and fermenting only the juice.

**Beer Ingredients:** Beer is slightly more complex than wine. Beer is mostly water, about 90 percent depending on the style and brand. Barley is added to the water, which is the source of food for the yeast. Beer will typically be under 5 percent alcohol once it is finished. Beer brewers also add hops to the beer mixture to add a distinct, bitter flavor. Ale style beer has yeast that is fermented at the top of the mixture at a higher temperature while lager is fermented at the bottom of the mixture at a lower temperature.

## 7.5 LOUIS PASTEUR AND HIS PATENTED BEER MAKING EQUIPMENT

### 7.5.1 Germ Theory

In a simple experiment of drawing through an aspirator a current of outside air through a tube containing a little plug of cotton wool, Pasteur demonstrated that, as the current passed it deposited on this sort of filter some of the solid corpuscles contained in the air. After a year's study, Pasteur reached this conclusion : "Gases, fluids, electricity, magnetism, ozone, things known or things occult, there is nothing in the air that is conditional to life, except the germs that it carries." Pasteur was convinced that the atmosphere itself contained nothing which could produce life, but in fact life did exist within it.

### 7.5.2 Germ Theory and Beer

With a simple, yet insightful application of this discovery and the ground breaking science in Germ Theory of Disease and his Studies on Fermentation, Louis Pasteur revolutionized beer and wine making processes. He posited that the exposure of beer and wine to the open atmosphere would contaminate the substance, due to the living organisms in the air, responsible for the processes of fermentation. Therefore, in order to effectively protect a substance from infection, not only must it be properly heated to eliminate or "pasteurize" any living organisms already within the substance, but during and after the heating takes place, the substance must remain closed off from any atmospheric influences.

Previously, the wort was boiled and then exposed to the air for cooling, before the yeast was introduced to catalyze fermentation. Pasteur later developed and patented a device that would allow for the proper brewing of beer, always ensuring the wort stayed protected from outside influences, right up to the point of distribution. In Pasteur's process, the wort is kept in closed vessels and cooled by spraying the outside of the vessel with water. A special yeast was introduced into the mash after it cooled, thus preventing contamination of the wort with stray wild yeasts floating through the air.

### 7.5.3 The Pasteurization of Beer

During the terrible days of the supremacy of the Commune in Paris, at the end of the Franco-German war, Pasteur was occupied in the laboratory of M. Duclaux, at Clermont-Ferrand, in studying the diseases of beer, with a view to attempt to raise French beer to the higher standard of the German brewers. Beer is naturally more prone to disease than wine, on account of the comparatively large quantity of gummy and saccharine matters which it contains in a state favorable to rapid decay. When the fermentation of the wort of beer sets in at the high temperature to which it is raised in mashing the liquid requires to be rapidly cooled. So long as it remains between the temperatures of 77° and 95° of Fahrenheit's scale it is peculiarly liable to be attacked by the injurious ferments proper to acetic, lactic, and butyric acids. If the must of beer were spontaneously fermented, like the must of grape juice, an acid of putrid liquid would invariably be produced in the place of beer. In the old process of what was technically known as high fermentation, which is also the one that is still employed with the bitter beers and pale ales of England, the fermenting liquid was kept in barrels, at a temperature ranging from 64° to 68° Fahrenheit. In the process of "low fermentation" which is more generally employed by the brewers of Germany and

France, a slow fermentation is established at a lower temperature, during which the yeast settles down to the bottom of the tubs and casks.

### 7.5.4 Wine and Beer Yeast Fermentation Process

It really isn't important that you learn a lot about wine and beer making through the fermentation of yeast. It's a simple process. You really need to know only a few things to make good wine and beer.

Wine and beer making through yeast fermentation is a chemical process by which microorganisms such as bacteria, molds, and yeast break down sugars into water, alcohol, and carbon dioxide gas.

### 7.5.5 The Yeast Fermentation Process

Yeast fermentation has been around for a very long time. Often, when a person begins to research the fermentation process, they are not aware of the fact that this has been going on almost since all of recorded history. The idea of wine and beer making through yeast fermentation is something that has been going on for many many years, this is part of why it is so important to find out your information from the sources that can actually help you.

Normal Home Brewing Fermentation Process Is In Two Parts. Part 1 Aerobic (Oxygen is present) This is the initial rapid process where the yeast is doubling its colony size every 4 hours. (Usually 24-48 hours) Part 2 Anaerobic. (No oxygen present) Slower activity and the yeast focuses on converting sugar to alcohol rather that increasing the number of yeast cells.

### 7.5.6 Fermentation Basics How It Works

When yeast is used to ferment a solution containing a lot of sugar, fermentation usually stops although sugar and yeast are still present.

Yeast have evolved what some consider a "primitive" form of metabolism to use sugar for an energy source. In alcohol fermentation, yeasts that produce carbon dioxide and alcohol and energy as end products, when the alcohol reaches a certain level it will inhibit the yeast from fermenting any more sugar.

Yeast cells ferment sugar because they can! When there is no Oxygen available, only organism that can ferment are able to synthesize and reproduce. The alcohol and $CO_2$ increase in fermentation, eventually this inhibits the cells even though there is still sugar available. The level of alcohol tolerance by yeast varies from 5% to about 21% depending on yeast strain.

Brewer's yeast can tolerate up to about 5% alcohol. Anything beyond this alcohol level the yeast cannot continue the fermentation process. Wine yeast on the other hand tolerates up to about 12% alcohol.

The fermentation process has other limits such as temperature. Greater than 27C kills the yeast less and than 15C results in yeast activity which is too slow. The amount of sugar in the solution can be too high and this can prevent fermentation. Some recipes suggest adding the sugar in parts throughout fermentation rather that all at the beginning. This is especially true if the brew is aimed at producing a high level of alcohol. Some yeast strains have evolved to handle higher sugar levels. Yeast such as Tokay and Sauterne.

## 7.6 MAKING HOMEMADE WINE AND BEER

Have you ever started a wine - and fermentation seems to start normally enough - but all of a sudden, the signs of fermentation seem to slow down or stop too soon? If so, you've experienced what's known as a "stuck fermentation."

### 7.6.1 Definition

By definition, a stuck fermentation is a fermentation that has stopped before all the available sugar in the wine has been converted to alcohol and $CO_2$. Were you to give up on the wine at this point, it would taste semi-sweet and pretty bad. That would be a shame, and what's more, a waste of good juice!

How did this situation occur? More importantly, what can you do to restart fermentation and salvage your wine?

### 7.6.2 Is it Really Stuck?

Before we dive into these questions, we should first make sure that our wine is stuck. Ask yourself these questions before you start dumping yeasts, additives, and chemicals willy-nilly into your carboy:

1. What is the SG (specific gravity) of your wine? Do you have proof that the SG is no longer falling, or is tremendously sluggish? If you don't know (or aren't sure how to do this), we recommend you see our article on how to use the hydrometer. It explains the ins and outs of measuring your wine's SG.
2. Do you have a good airtight seal at your airlock? Is your airlock firmly seated in the bung, and is the bung securely seated in the mouth of the carboy? If not, this might explain why you don't see bubbles in your airlock.
3. Are you fermenting in hot weather or in a hot area? Yeast works faster under higher (yet tolerable) temperatures, so your wine may actually be finished fermenting before you realize it.

Luckily, stuck fermentations don't occur very often - but when they do, it's important to make corrections right away and get the fermentation going again.

### 7.6.3 Causes of Stuck Fermentations

More than likely, the cause of a stuck fermentation centers around the wine yeast. Either something in the wine environment is preventing the yeast from working properly, or there is a problem with the yeast itself.

Even if the proper yeast is used, most experienced vintners know that wine yeast is pretty particular when it comes to fermenting wine to dryness - the proper environmental conditions must be met, and nutrients need to be available for the yeast to continue their hard work.

Wine yeast is most happy when:

- It's not too hot, and not too cold
- There's lots of food to eat
- No killer agents are present
- They live in sanitary conditions
- Oxygen is available (to kick off fermentation)

Using a little common sense, then (which vintners seem to have a lot of!), we can easily extrapolate the major causes of a stuck fermentation:

1. Extreme fermentation temperatures - too high or too low
2. Using unsanitized equipment - dirty or unsanitary equipment increases the possibility that microbiological factors such as wild killer yeasts and bacteria will spoil your wine
3. Using old yeast - weakened or expired/out-of-date
4. Incorrect yeast used - match the proper yeast for your wine (in the case of buying wine concentrate kits from grapestompers, this is automatically done for you)
5. Yeast not rehydrated before pitching - always rehydrate yeast according to manufacturer's suggestions
6. Yeast rehydrated at too low or too high a temperature - this can kill a large percentage of yeast cell population
7. Temperature shock when rehydrated yeast is introduced to must - try to allow no more than a 5-7° C differential between yeast mixture and must
8. Sulfite levels too high - adding too much metabisulfite; failing to wait 24 hours after Campden applied to must before pitching yeast; or high must pH, which can lead to high fermentation rate
9. Pesticide residue on the exterior surface of grapes or fruits - wash all grapes or fruits well before processing
10. Lack of nutrients, including a lack of nitrogen or certain amino acids
11. Extremely high starting SG - too much sugar in must at the outset
12. Sugar has all been utilized - you don't want your starting SG to be too low either!

13. Too much $CO_2$ in your wine - don't forget to degas
14. Naturally occurring sorbate in must - as in the case of blueberries

## 7.7 PREVENTION OF STUCK FERMENTATIONS

Here are some things the home winemaker can do to prevent stuck fermentations:

- Monitor and ensure proper fermentation temperatures
- Ensure proper sanitation - learn how to sanitize equipment
- Use fresh yeast
- Use the proper yeast for the wine you're making - don't guess or use a packet of yeast just because it's handy
- Properly rehydrate yeast before pitching
- Pitch the yeast within 20 minutes of rehydrating it
- Maintain proper free $SO_2$ levels - the amount of metabisulfite to add to your wine depends on pH of wine
- Add yeast nutrient before pitching yeast
- Keep your starting SG to reasonable levels (1.090 - 1.100 or lower). If you don't currently have a hydrometer, buy one and learn how to use it.
- Aerate the must properly by vigorous stirring, just before pitching the yeast. This will introduce the oxygen needed to "kick off" fermentation.

### 7.7.1 Treatment of Stuck Fermentations

And here's what to do if you get stuck... and remember - always start with the simplest things first. Resist the urge to add yeast or additives until you've tried the easy things.

1. Adjust the temperature of your wine. In most cases we've seen, simply warming your wine to 70-75° F for a couple of days will get the ball rolling.
2. Rouse the yeast by swishing or stirring the lees (trub) - sometimes moving the yeast around in the wine will get fermentation going again.
3. Warning: Although it may be tempting, don't add vitamins (yeast nutrient) during stuck fermentations. Leftover vitamins can stimulate spoilage microbes. Only add a yeast nutrient before or as you pitch your yeast. If you want to add a yeast energizerat this point, that's OK. Simply go to the local drug store and ask the pharmacist for some Thiamin HCL. Add 25 mg. per gallon of wine and mix well.
4. Remove the old yeast by racking the wine, then re-inoculate with fresh yeast, preferably a killer strain like Lalvin EC-1118 or Red Star Premier Cuvee. In a

pinch, you could even use a Red Star Champagne yeast. We want to get rid of the old yeast because yeast cells seem able to detect the presence of other dying cells, and are more likely to get "lazy" themselves.

5. If you detect there is a nitrogen deficiency (less than 200 mg/L fermentable nitrogen), addition of DAP (diammonium phosphate dibasic - commonly known in the winemaking industry as Fermaid) is called for.

If none of the above seem to help restart your fermentation within a couple or three days, it's time to bring in the heavy hitters:

1. Make a yeast starter by pulling off approximately 1/2 gallon of must, and add 1.5 to 2 teaspoons of yeast energizer (thiamin HCL) and 1 packet of "killer" or champagne yeast. Mix well, cover loosely and place in a warm spot. Once you have a vigorous fermentation you can add it back to the original must. OR...
2. Make a different kind of starter: use about a 1/2 cup of warm water, dissolve 1 teaspoon of sugar in the water, add some orange juice to this mix, make sure the temperature is about 90° F, before adding a packet of Red Star Premier Cuvee or Lalvin EC-1118 yeast to this mixture. Wait until it really gets working. Take about a gallon of your must and warm it up to about 68° to 70° F. Now add the yeast starter to the gallon of must, as it starts to work and gets going, SLOWLY add small portions of the stuck fermentation to that which is working. You should not add more than a quart, make sure the temperature of that which you are adding is at least 70° F. As the volume of the working must gets larger, you can add larger portions to the fermentation. Make sure the temperature is at least 70° F before you add it.

If you are interested in wine making then you need to do a little homework before you get started. Wine making is not something that you can just plunge into and learn along the way, you need to do some research so that you know that you are not missing any vital wine making steps or processes. If you do miss these wine making your wine could very well turn out tasting like anything but wine.

Wine making is a something fun that you can do in your spare time and at the end you will have a glorious result. Wine making is something that people tend to get hooked on because the very first batch is not usually perfect, it may taste delicious but it will not be perfect. As you learn about wine making and as you get the wine making experience behind you your wine will get better and better. You will undoubtedly pick up tips and tricks to help you improve your wine making capabilities, whether they come from practice, talking with fellow wine makers, or reading informative books on the subject.

The vast majority of wines use the grape for a reasons and this is because this little grape will have so many of the necessary wine ingredients already infused into it. In wine making your success is going to be all about balance and chemicals and the grape has many of the necessary chemicals in it already.

For good and delicious wine making you need to have the right balance of sugar, tannin, moisture and nutrients, well, it just so happens that the grape fits the bill to perfection. This allows for the grape to ferment, as it needs to for successful wine making. Choosing the right grape is only the beginning of your wine making journey though.

There are many different grapes that you can use and when you get more skilled at wine making you might want to add a little of some other fruits to the mix to give your wine some extra depth and flavoring. Read about the different grapes that can be used in wine making and how they affect the outcome of the wine making process. You know what kind of wine you like so this will give you an idea of which direction to take your wine making in.

Wine making is fun, remember that as you go along. If you make any mistakes along the wine making way don't worry about it, you are new and it happens to everyone once and awhile.

While beer is widely available mostly anywhere and there are thousands of varieties to choosefrom, making your own beer can be an exciting hobby. Your quest to look for the right kind of beer for your taste might be such a confusing journey as there are thousands of varieties out there. Perhaps making your own beer is the good way to speed up your pursuit. Brewing beer is simple if you follow every step religiously. Here's an overview of how to do it.

The main ingredient for beer is barley, which is the source of sugar to be fermented for the brewing process. Rice, corn, and what are sometimes used as supplementary ingredients. To prepare for the beer making process, barley is malted by steeping its grains in water until they begin to sprout. At a certain point the grains are dried to stop the germination process and the resulting product from malting is the grain bill. The grain bills are then mashed in a special mill that crushes the grain's starch center but still keeping the husk whole. This is necessary for the proper change of starch to become sugar.

The next step is sparging, and this is done by rinsing the mashed grain bills to separate the husk from the sugar. In a brew kettle, water is slowly trickled in to the bed of mashed grain. Sweet liquor would be drained from the brew kettle and collected. What is collected are known as malt extract. Those who are in a hurry could actually buy concentrated malt and start off with that, to make life easier. Malt extracts are classified according to their origin, color and the capacity for fermentation.

One important element to making bear is yeast. Without it, beer cannot be brewed. A special type of yeast called brewer's yeast is specifically used for making beer. Brewer's yeast is even further classified into lager and ale yeasts. Lager yeast sinks during the fermentation process and requires low temperatures of about 40 to 50 degrees Fahrenheit.

Ale yeast on the other hand remains on top during the fermentation process and requires temperatures of about 55 to 65 degrees F. Ale yeast is used more often for home brewing because it is difficult to maintain the temperature required by lager yeast.

To brew beer, malted grain is steeped into the brewing water using a steeping bag which is like a giant tea bag. The brewing pot is then heated until right before the water boils, at this point the steeping bag is removed. Once the water is boiling well, the heat is turned off. Malt extract is then mixed into the water until the ingredients are incorporated very well. The pot will be heated once more until the mixture, called wort, boils again.

The next step is called hopping. This involves infusing bittering flowers into the wort by means of hopping bags. The bittering hop bag is just placed into the brew pot and boiled for about an hour to extract the flavors of the hops. As the wort boils, certain proteins will form. Towards the end of boiling, around fifteen minutes before turning the heat off, a teaspoon of Irish Moss should be added to the wort to remove proteins that could make the beer hazy. Towards the end of the wort's boiling process, finishing hops are added for more aroma and flavor.

Once boiling is done, the hops bag is removed. The brew pot is then covered tightly with an aluminum foil sheet. The wort is then chilled to cool it enough for the yeast by placing the brew pot in a cold bath. The wort should be cooled as quick as possible to avoid prolonged exposure to possible contaminants. The wort would then be aerated to allow fermentation process. The wort is poured into a sanitized fermenter and allowed to splash and form. It is important not to pour everything as there are sediments at the bottom of brew pot that should not be mixed to the final brew. Some cold water is added on top of the wort.

The next step is called pitching. This is done when the fermenter has cooled the wort to about less then 75 degrees F. Yeast would then be sprinkled on top of the wort. It is important to be sanitize everything used for this step, even the scissors used to open the yeast's packet. After pouring the yeast, the fermenter would then be sealed using an air lock. The fermentation process can take from about two to five days. It is important to check the fermentation daily. Fermentation is done when the bubble rate in the air lock becomes less than a time per two minutes, it takes about a week to reach this point. After fermentation, the beer would be bottled in a special process. Afterwards, the beer would be aged by storing the beer in a dark place for two weeks at room temperature, and for another two weeks at cool temperature. Once this is done, the beer may then be chilled for 24 hours, and then it can enjoyed.

## 7.8 CONCLUSION

Fermentation is a chemical change brought on by the action of micro scopic yeast, mol ds and bacteria. The souring of milk, the rising of dough and the conversion of sugar to alcohol are all examples of fermentation. Of course the conversion of sugar to alcohol is what we are interested in so that is what we will be concentrating on.

Most wild yeast will give your beer or wine strange off-tastes and a lot of them are not very tolerable of alcohol which means you will end up with a partly fermented, low alcoholic, sweet beverage. That is why brewers and wine makers use yeast that is specifically cultured for the fermentation of alcoholic beverages.

# 8

# FERMENTATION OF DAIRY PRODUCTS

## 8.1 INTRODUCTION

Studies were conducted for retail or commercially manufactured yogurts and other cultured dairy products obtained in the U.S., Australia, Spain, France, Norway, Greece, Argentina, and South Africa. All of the yogurts examined contained the yogurt culture organisms, S. thermophilus and L. delbrueckii subsp. bulgaricus, at levels ranging from < 104 to 109 cfu/g or ml. In general, counts for S. thermophilus were somewhat higher than for L. delbrueckii subsp. bulgaricus. In several studies, other microorganisms, including Bifidobacterium spp. and Lactobacillus spp., were also enumerated. Levels of the latter ranged from undetectable (< 10 cfu/g) to 108 cfu/g. The addition of these probiotic bacteria did not appear to have any effect on levels of the yogurt culture organisms. Although most studies reported counts at only a single time point, other studies reported initial counts as well as at a second time point, usually considered end-of-shelf-life. In such cases, counts were generally similar at both time points (>106 cfu/g), provided samples were stored at refrigeration temperatures

In addition to fresh yogurt, frozen yogurt was also examined for bacteria. Results from several studies indicates that when these products were assessed for the relevant yogurt LAB, levels were generally similar to fresh yogurt, with counts ranging from 104 to 109 cfu/g. The stability of lactic cultures in frozen yogurt during long-term storage at freezer temperature (-23°C) has also been studied. In general, LAB (S. thermophilus and L. delbrueckii subsp. bulgaricus) survived beyond the designated shelf-life period (1 year), with less than a 0.5 log reduction for most samples.

The number and type of live microorganisms in other cultured dairy products have also been reported. These include kefir, cultured buttermilk and simply "fermented milk." As for other cultured dairy products, populations of LAB were in the 105–109 cfu/g range.

### 8.2.1 History of Dairy Products

There is evidence of cultured milk products in cultures as far back as 2000 BCE. The earliest yoghurt was probably fermented spontaneously, perhaps by wild bacteria residing inside goatskin bags used for transportation.

In the records of the ancient culture of Indo-Iranians (Iran and India), yoghurt is mentioned by 500 BCE. In this record the combination of yoghurt and honey is called "the food of the gods". Persian traditions hold that "Abraham owed his fecundity and longevity to the regular ingestion of yogurt".

The oldest writings mentioning yoghurt are attributed to Pliny the Elder, who remarked that certain nomadic tribes knew how "to thicken the milk into a substance with an agreeable acidity". The use of yoghurt by medieval Turks is recorded in the books Diwan Lughat al-Turk by Mahmud Kashgari andKutadgu Bilig by Yusuf Has Hajib written in the 11th century. Both texts mention the word "yoghurt" in different sections and describe its use by nomadic Turks. An early account of a European encounter with yoghurt occurs in French clinical history: Francis I suffered from a severediarrhoea which no French doctor could cure. His ally Suleiman the Magnificent sent a doctor, who allegedly cured the patient with yoghurt. Being grateful, the French king spread around the information about the food which had cured him.

**Fig. 1. Raita is a condiment made with yoghurt and popular in India and Pakistan.**

Until the 1900s, yoghurt was a staple in diets of people in the Russian Empire (and especially Central Asia and theCaucasus), Western Asia, South Eastern Europe/Balkans, Central Europe, and India. Stamen Grigorov (1878–1945), a Bulgarian student of medicine in Geneva, first examined the microflora of the Bulgarian yoghurt. In 1905, he described it as consisting of a spherical and a rod-like lactic acid bacteria. In 1907 the rod-like bacteria was calledLactobacillus bulgaricus (now Lactobacillus delbrueckii subsp. bulgaricus). The Russian Nobel laureate biologistIlya Ilyich Mechnikov, from the

Institut Pasteur in Paris, was influenced by Grigorov's work and hypothesised that regular consumption of yoghurt was responsible for the unusually long lifespans of Bulgarian peasants. BelievingLactobacillus to be essential for good health, Mechnikov worked to popularise yoghurt as a foodstuff throughout Europe.

Isaac Carasso industrialised the production of yoghurt. In 1919, Carasso, who was from OttomanSalonika, started a small yoghurt business in Barcelona, Spain, and named the business Danone("little Daniel") after his son. The brand later expanded to the United States under an Americanised version of the name: Dannon.

**Fig. 2. Tarator is a cold soup made of yoghurt and cucumber (dill, garlic, walnuts and sunflower oil are sometimes added) and is popular in Bulgaria.**

Yoghurt with added fruit jam was patented in 1933 by the Radlická Mlékárna dairy in Prague. It was introduced to the United States in 1947, by Dannon.

Yoghurt was first introduced to the United States in the first decade of the twentieth century, influenced by Élie Metchnikoff's The Prolongation of Life; Optimistic Studies; it was available in tablet form for those with digestive intolerance and for home culturing. It was popularised by John Harvey Kellogg at the Battle Creek Sanitarium, where it was used both orally and in enemas, and later by Armenian immigrants Sarkis and Rose Colombosian, who started "Colombo and Sons Creamery" in Andover, Massachusetts in 1929. Colombo Yoghurt was originally delivered around New England in a horse-drawn wagon inscribed with the Armenian word "madzoon" which was later changed to "yogurt", the Turkish name of the product, as Turkish was the lingua franca between immigrants of the various Near Easternethnicities who were the main consumers at that time. Yoghurt's popularity in the United States was enhanced in the 1950s and 1960s, when it was presented as a health food. By the late 20th century yoghurt had become a common American food item and Colombo Yogurt was sold in 1993 to General Mills, which discontinued the brand in 2010.

## 8.3 NUTRITIONAL VALUE AND HEALTH BENEFITS

Yoghurt is nutritionally rich in protein, calcium, riboflavin, vitamin B6 and vitamin B12. It has nutritional benefits beyond those of milk. People who are moderately lactose-intolerant can consume yoghurt without ill effects, because much of the lactose in the milk precursor is converted to lactic acid by the bacterial culture.

Yoghurt containing live cultures is sometimes used in an attempt to prevent antibiotic-associated diarrhea.

A study published in the International Journal of Obesity (11 January 2005) also found that the consumption of low-fat yoghurt can promote weight loss, especially due to thecalcium in the yoghurt.

### 8.3.1 Varieties and Presentation

Yoghurt is popular in Nepal, where it is served as both an appetiser and dessert. Locally called dahi, it is a part of the Nepali culture, used in local festivals, marriage ceremonies, parties, religious occasions, family gatherings, and so on. The most famous type of Nepalese yoghurt is called juju dhau, originating from the city of Bhaktapur. In Tibet, yak milk is made into yoghurt and consumed.

In Northern Iran, Mâst Chekide is a variety of kefir yoghurt with a distinct sour taste. It is usually mixed with a pesto-like water and fresh herb purée called delal. Yoghurt is a side dish to all Iranian meals. The most popular appetisers are spinach or eggplant borani, Mâst-o-Khiâr with cucumber, spring onions and herbs, and Mâst-Musir with wild shallots. In the summertime, yoghurt and ice cubes are mixed together with cucumbers, raisins, salt, pepper and onions and topped with some croutons made of Persian traditional bread and served as a cold soup. Ashe-Mâst is a warm yoghurt soup with fresh herbs, spinach and lentils. Even the leftover water extracted when straining yoghurt is cooked to make a sour cream sauce called kashk, which is usually used as a topping on soups and stews.

Matsoni is a Georgian yoghurt popular in the Caucasus and Russia. It is used in a wide variety of Georgian dishes and is believed to have contributed to the high life expectancy and longevity in the country. Dannon used this theory in their 1978 TV advertisement called In Soviet Georgia where shots of elderly Georgian farmers were interspersed with an off-camera announcer intoning, "In Soviet Georgia, where they eat a lot of yogurt, a lot of people live past 100." It is made with Lactobacillus delbrueckii subsp. bulgaricus and Streptococcus thermophilus lactic acid bacteria. Matsoni is also popular in Japan under the name Caspian Sea Yogurt.

Tarator and Cacık are popular cold soups made from yoghurt, popular during summertime in Albania, Bulgaria, Republic of Macedonia, Serbia and Turkey. They are made with ayran, cucumbers, dill, salt, olive oil, and optionally garlic and ground walnuts. Tzatziki in Greece and milk salad in Bulgaria are thick yoghurt-based salads similar to tarator.

Rahmjoghurt, a creamy yoghurt with much higher fat content (10%) than most yoghurts offered in English-speaking countries, is available in Germany and other countries.

Cream-top yoghurt is yoghurt made with unhomogenised milk. A layer of cream rises to the top, forming a rich yoghurt cream. Cream-top yoghurt was first made commercially popular in the United States by Brown Cow of Newfield, New York, bucking the trend toward low- and non-fat yoghurts.

Raita is a yoghurt-based South Asian/Indian condiment, used as a side dish. The yoghurt is seasoned with cilantro (coriander), cumin, mint, cayenne pepper, and other herbs and spices. Vegetables such as cucumber and onions are mixed in, and the mixture is served chilled. Raita has a cooling effect on the palate which makes it a good foil for spicy Indian dishes.

Dudh is a Sindhi-curd, popular in India. People drink dudh along with food at intervals, to help digestion and make food more delicious. In some places dudh is also served with plain rice.

Dahi is a yoghurt of the Indian subcontinent, known for its characteristic taste and consistency. The word dahi seems to be derived from the Sanskrit word dadhi, one of the five elixirs, or panchamrita, often used in Hindu ritual. Dahi also holds cultural symbolism in many homes in the Mithilanchalregion of Bihar. It is found in different flavours, two of which are famous: sour yoghurt (tauk doi) and sweet yoghurt (meesti or podi doi). In India, it is often used in cosmetics mixed with turmeric and honey. Sour yoghurt is also used as a hair conditioner by women in many parts of India.Dahi is also known as Thayiru (Malayalam), doi (Assamese, Bengali), dohi (Oriya), perugu (Telugu),Mosaru (Kannada), Thayir (Tamil), or Qəzana a pəəner (Pashto).

Srikhand, a popular dessert in India, is made from drained yoghurt, saffron, cardamom, nutmeg and sugar and sometimes fruits such as mango or pineapple.

## 8.3 SWEETENED AND FLAVOURED YOGHURT

To offset its natural sourness, yoghurt can be sold sweetened, flavoured or in containers with fruit or fruit jam on the bottom. If the fruit has been stirred into the yoghurt before purchase, it is commonly referred to in the United States as Swiss-style. Most yoghurts in North America have added pectin, found naturally in fruit, and/or gelatin to create thickness and creaminess artificially at lower cost. This type of yoghurt is also marketed under the name Swiss-style, although it is unrelated to the way yoghurt is eaten in Switzerland. Some yoghurts, often called "cream line," are made with whole milk which has not been homogenised so the cream rises to the top. Fruit jam is used instead of raw fruit pieces in fruit yoghurts to allow storage for weeks.

Sweeteners such as cane sugar or sucralose – for low-calorie yoghurts – are often present in large amounts in commercial yoghurt.

In the UK and USA, sweetened, flavoured yoghurt is the most popular type, typically sold in single-serving plastic cups. Typical flavours are vanilla, honey, or fruit such as strawberry, blueberry,blackberry, raspberry, mango or peach. In recent years, some manufacturers are marketing flavours inspired by desserts, such as chocolate or cheesecake, with many variants.

In Australia, flavoured and Greek are the two most popular types of yoghurt, and are usually sold in one kilogram tubs.

### 8.3.1 Home Preparation

Yoghurt can be easily prepared at home. To produce a quart of yoghurt, a quart of milk is heated (generally to 180 °F or 80 °C) and then cooled to approximately 110 °F or 43 °C. An inexact test for the temperature is to test it with a finger: if the finger can be left in the milk for ten seconds, the milk is at about the correct temperature. This does risk contaminating the yoghurt with bacteria under the fingernails, though.

Once the milk has cooled, a tablespoon of yoghurt containing live active cultures is added to the scalded milk and mixed well. The yoghurt is poured into a container and incubated for the desired length of time. The fermentation time depends on the desired texture and tang in the final product. A shorter period yields a slightly runnier product with a mildly sour flavor, while a longer period yields a thicker and tarter yoghurt.

To store plain yoghurt for the purpose of culturing future batches, it can be frozen in one tablespoon portions and then stored in the freezer. The culture can be thawed in the refrigerator or at room temperature and then used.

## 8.4 STRAINED YOGHURTS

Strained yoghurts are types of yoghurt which are strained through a paper or cloth filter, traditionally made of muslin, to remove the whey, giving a much thicker consistency and a distinctive, slightly tangy taste. Strained yoghurt is becoming more popular with those who make yoghurt at home, especially if using skim milk which results in a thinner consistency. Once yoghurt is made and refrigerated overnight, it is poured in a muslin or cheesecloth bag and hung in the coolest place in the house, with a tub placed underneath to collect the dripping whey. In cold weather a single day (or night) of straining is sufficient. In higher ambient temperatures yoghurt will spoil rapidly, therefore it had best be actively squeezed or strained until about a third or more of its initial weight has run off. The remainder is now strained and is refrigerated again.

Labneh is a strained yoghurt used for sandwiches popular in Arab countries. Olive oil, cucumber slices, olives, and various green herbs may be added. It can be thickened further and rolled into balls, preserved in olive oil, and fermented for a few more weeks. It is sometimes used with onions, meat, and nuts as a stuffing for a variety of pies or kebbeh balls.

Some types of strained yoghurts are boiled in open vats first, so that the liquid content is reduced. The popular East Indian dessert, a variation of traditional dahi called mishti dahi, offers a thicker, more custard-like consistency, and is usually sweeter than western yoghurts.

Strained yoghurt is also enjoyed in Greece and is the main component of tzatziki, a well-known accompaniment to gyros and souvlaki pita sandwiches: it is a yoghurt sauce or dip made with the addition of grated cucumber, olive oil, salt and, optionally, mashed garlic.

## 8.5 HOW PRODUCTS ARE MADE?

Yogurt is a dairy product, which is made by blending fermented milk with various ingredients that provide flavor and color. Although accidentally invented thousands of years ago, yogurt has only recently gained popularity in the United States.

It is believed that yogurt originated in Mesopotamia thousands of years ago. Evidence has shown that these people had domesticated goats and sheep around 5000 B.C. The milk from these animals was stored in gourds, and in the warm climate it naturally formed a curd. This curd was an early form of yogurt. Eventually, a process for purposely producing yogurt was developed.

While yogurt has been around for many years, it is only recently (within the last 30-40 years) that it has become popular. This is due to many factors including the introduction of fruit and other flavorings into yogurt, the convenience of it as a ready-made break-fast food and the image of yogurt as a low fat healthy food.

Manufacturers have responded to the growth in the yogurt market by introducing many different types of yogurt including low fat and no-fat, creamy, drinking, bio-yogurt, organic, baby, and frozen. Traditional yogurt is thick and creamy. It is sold plain and in a wide assortment of flavors. These are typically fruit flavors such as strawberry or blue-berry however, newer, more unique flavors such as cream pie and chocolate have also been introduced. Cereals and nuts are some-times added to yogurts. Yogurt makers also sell products with a varying level of fat. Low fat yogurt, which contains between 0.5% and 4% fat, is currently the best selling. Diet no-fat yogurt contains no fat at all. It also contains artificial sweeteners that provide sweetness while still reducing calories. Creamy yogurt is extra thick, made with whole milk and added cream. Drinking yogurt is a thinner product, which has a lower solids level than typical yogurt. Bio-yogurt is made with a different type of fermentation culture and is said to aid digestion. Yogurt that is made with milk from specially fed cows is called organic yogurt. This type of yogurt is claimed to be more nutritious than other yogurts. Other types of yogurts include pasteurized stirred yogurt that has extended shelf life, baby yogurt made specifically for children, and frozen yogurt.

The yogurt itself has a generally aldehydic flavor, which is a result of the fermentation process. Since it is made from milk, yogurt is rich in nutrients. It contains protein and vitamins and is a rich source of calcium. In fact, a small container of yogurt contains as much calcium as a third of a pint of milk. In addition to these nutritional characteristics, yogurt is also thought to have additional health benefits. One of the suggested benefits of yogurt is that it acts as a digestive aid. In the body, it is thought that yogurt can encourage the growth of beneficial bacteria in the gut. These organisms help to digest food more efficiently and protect against other, harmful organisms. Another health benefit of yogurt is for people that are lactose intolerant. These people have difficulty digesting milk products however, they typically can tolerate yogurt.

### 8.5.1 Raw Materials

In general, yogurt is made with a variety of ingredients including milk, sugars, stabilizers, fruits and flavors, and a bacterial culture. During fermentation, these organisms interact with the milk and convert it into a curd. They also change the flavor of the milk giving it the characteristic yogurt flavor of which acetaldehyde is one of the important contributors. The primary byproduct of the fermentation process is lactic acid. The acid level is used to determine when the yogurt fermentation is completed which is usually three to four hours. The suppliers of these yogurt cultures offer various combinations of the two bacterial types to produce yogurts with different flavors and textures.

To modify certain properties of the yogurt, various ingredients may be added. To make yogurt sweeter, sucrose (sugar) may be added at approximately 7%. For reduced calorie yogurts, artificial sweeteners such as aspartame or saccharin are used. Cream may be added to provide a smoother texture. The consistency and shelf stability of the yogurt can be improved by the inclusion of stabilizers such as food starch, gelatin, locust-bean gum, guar gum and pectin. These materials are used because they do not have a significant impact on the final flavor. The use of stabilizers is not required however, and some marketers choose not to use them in order to retain a more natural image for their yogurt.

To improve taste and provide a variety of flavors, many kinds of fruits are added to yogurt. Popular fruits include strawberries, blueberries, bananas, and peaches, but almost any fruit can be added. Beyond fruits, other flavorings are also added. These can include such things as vanilla, chocolate, coffee, and even mint. Recently, manufacturers have become quite creative in the types of yogurt they produce using natural and artificial flavorings.

### 8.5.2 The Manufacturing Process

The general process of making yogurt includes modifying the composition of and pasteurizing the milk; fermenting at warm temperatures; cooling it; and adding fruit, sugar, and other materials.

### 8.5.3 Modifying Milk Composition

- When the milk arrives at the plant, its composition is modified before it is used to make yogurt. This standardization process typically involves reducing the fat content and increasing the total solids. The fat content is reduced by using a standardizing clarifier and a separator (a device that relies upon centrifugation to separate fat from milk). From the clarifier, the milk is placed in a storage tank and tested for fat and solids content. For yogurt manufacture, the solids content of the milk is increased to 16% with 1-5% being fat and 11-14% being solids-not-fat (SNF). This is accomplished either by evaporating off some of the water, or adding concentrated milk or milk powder. Increasing the solids content improves the nutritional value of the yogurt, makes it easier to produce a firmer yogurt and improves the stability of the yogurt by reducing the tendency for it to separate on storage.

### 8.5.4 Pasteurization and Homogenization

- After the solids composition is adjusted, stabilizers are added and the milk is pasteurized. This step has many benefits. First, it will destroy all the microorganisms in the milk that may interfere with the controlled fermentation process. Second, it will denature the whey proteins in the milk which will give the final yogurt product better body and texture. Third, it will not greatly alter the flavor of the milk. Finally, it helps release the compounds in milk that will stimulate the growth of the starter culture. Pasteurization can be a continuous-or batch-process. Both of these processes involve heating the milk to a relatively high temperature and holding it there for a set amount of time. One specific method for batch process pasteurization is to heat a large, stainless steel vat of milk to 185° F (85° C) and hold it there for at least 30 minutes.
- While the milk is being heat treated, it is also homogenized. Homogenization is a process in which the fat globules in milk are broken up into smaller, more consistently dispersed particles. This produces a much smoother and creamier end product. In commercial yogurt making, homogenization has the benefits of giving a uniform product, which will not separate. Homogenization is accomplished using a homogenizer or viscolizer. In this machine, the milk is forced through small openings at a high pressure and fat globules are broken up due to shearing forces.

### 8.5.5 Fermentation

- When pasteurization and homogenization are complete, the milk is cooled to between 109.4-114.8° F (43-46° C) and the fermentation culture is added in a concentration of about 2%. It is held at this temperature for about three to four hours while the incubation process takes place. During this time, the bacteria metabolizes certain compounds in the milk producing the characteristic yogurt flavor. An important byproduct of this process is lactic acid.

- Depending on the type of yogurt, the incubation process is done either in a large tank of several hundred gallons or in the final individual containers. Stirred yogurt is fermented in bulk and then poured into the final selling containers. Set yogurt, also known as French style, is allowed to ferment right in the container it is sold in. In both instances, the lactic acid level is used to determine when the yogurt is ready. The acid level is found by taking a sample of the product and titrating it with sodium hydroxide. A value of at least 0.9% acidity and a pH of about 4.4 are the current minimum standards for yogurt manufacture in the United States. When the yogurt reaches the desired acid level, it is cooled, modified as necessary and dispensed into containers.

### 8.5.6 Controlling Yoghurt Fermentation

The most important component in controlling the quality of yogurt fermentation is temperature. Temperature affects the yogurt fermentation by:

**Controlling the growth rate of the microorganisms:** If the temperature is too low, the culture grows too slowly to adequately acidify milk and to achieve a good texture. The commercial starter is a mixed culture of thermophilus and L. bulgaricus. If the temperature is too high it might end up killing the cultures.

Temperature will affect the taste of the yogurt produced a the formation and secretion of metabolites which contribute to the overall taste are dependent on the growth rate.

The temperature range of proper yogurt fermentation is quite small, i.e. from 42 °C to 44 °C. Higher temperature tends to give a sweeter yogurt as the rate of metabolism is higher. It will also makes the yogurt set faster.

It is very important that once the desired acidity is reached the fermentation is halted by low temperature. This cooling step is of most importance in the industrial production of yogurt, if not the taste of yogurt will be affected.

### 8.5.7 Adding Other Ingredients

- Fruits, flavors, and other additives can be added to the yogurt at various points in manufacturing process. This is typically dependent on the type of yogurt being produced. Flavor in non-fruit yogurts are added to the process milk before being dispensed into cartons. Fruits and flavors can also be added to the containers first, creating a bottom layer. The inoculated milk is then added on top and the carton is sealed and incubated. If the fruit is pasteurized, it can be added as a puree to the bulk yogurt, which is then dispensed into containers. Finally, the fruit can be put into a special package, which is mixed with plain yogurt upon consumption.
- The finished yogurt containers are placed in cardboard cases, stacked on pallets, and delivered to stores via refrigerated trucks.

### 8.5.8 Quality Control

Milk products such as yogurt are subject to a variety of safety testing. Some of these include tests for microbial quality, degree of pasteurization, and various forms of contaminants. The microbial quality of the incoming milk is determined by using a dye reaction test. This method shows the number of organisms present in the incoming milk. If the microbial count is too high at this point, the milk may not be used for manufacture. Since complete pasteurization inactivates most organisms in milk, the degree of pasteurization is determined by measuring the level of an enzyme in the milk called phosphatase. Governmental regulations require that this test be run to ensure that pasteurization is done properly. Beyond microbial contamination, raw milk is subject to other kinds of contaminants such as antibiotics, pesticides or even radioactivity. These can all be found through safety testing and the milk is treated accordingly.

In addition to safety tests, the final yogurt product is also evaluated to ensure that it meets the specifications set by the manufacturer for characteristics such as pH, rheology, taste, color, and odor. These factors are tested using various laboratory equipment such as pH meters and viscometers and also human panelists.

### 8.5.9 The Future

The future of yogurt manufacturing will focus on the development of new flavors and longer lasting yogurts. The introduction of new flavors will be driven by consumer desires and new developments by flavor manufacturers. The suppliers of the bacterial cultures are conducting research that hints at the development of uniquely flavored yogurts. By varying the types of organisms in the cultures, yogurt is produced much faster and lasts longer than conventional yogurt.

Additionally, the nutritional aspects of yogurt will be more thoroughly investigated There is some evidence that has shown consumption of yogurt has a beneficial antibiotic effect. It has also been shown to reduce the incidence of lactose intolerance and other gastro-intestinal illnesses. Other purported benefits of yogurt include the reduction of cholesterol, protection against certain cancers, and even boosting the immune system. The research is still not complete on these benefits however, these factors will likely be important in the continued market growth of yogurt.

## 8.6 CHEESE

Although considerable microbiological data for cheese exists, most of these reports are concerned with microorganisms having public health or cheese quality implications. Still, levels of lactic acid and related bacteria were reported for more than 30 types of cheese from 18 countries including the United States, Italy, France, Germany, Mexico, Ireland, and South Africa. Many papers reported the microorganisms as mesophilic streptococci, lactococci, and lactobacilli or as thermophilic streptococci and lactobacilli. Others reported total microorganisms and total LAB. For most products, only one time period was recorded (usually the most aged sample). Microbial counts ranged from undetectable (< $10^3$ cfu/g) to $10^9$ cfu/g, with the highest levels found in Tilsit cheese (typically aged 2–4 months). In contrast, Grana Padano aged 1 year, Parmesan aged greater than 1 year, and Swiss Gruyere aged greater than 1 year all showed no detectable microorganisms (< $10^3$ cfu/g). As noted for other products, the methods used by the investigators may have influenced the reported data. Thus, enumeration of selected organisms (e.g., S. thermophilus) was only possible if the appropriate medium and growth conditions were used.

## 8.7 FERMENTED MEATS

In general, samples were either obtained from retail, directly from manufacturers, or were produced via industrial conditions. Most samples were from the United States, Spain, Portugal, and Italy and were composed of pork and/or beef. The levels of microorganisms (LAB and total) ranged from undetectable (< $10^2$ cfu/g) to $10^{10}$ cfu/g. Data were reported as either within the product shelf life or after ripening or maturation of the sausage. Counts of viable microorganisms in sausages from the United States were generally lower (< $10^7$ cfu/g) compared to sausages from other countries. In particular, LAB levels were all < $10^6$ cfu/g. In contrast, several of the European sausages contained high levels of LAB (>$10^8$ cfu/g.). European sausages were more often artisan sausages from smaller manufacturers, although similar microorganisms are used in comparison to sausages from the United States.

## 8.8 FERMENTED VEGETABLES

Fermented cucumbers products were also considered. Laboratory-manufactured products, using industrial or traditional practices, were included due to the lack of literature on fermented vegetables from retail sources.

Microbial counts for sauerkraut were generally reported as LAB with counts ranging from $10^3$ to $10^8$ cfu/g. Reported samples were for sauerkraut originating from the United States, Finland, and Croatia. Levels of LAB and Lactobacillus were reported for olives produced in Italy, Greece, Portugal, Spain, and the United States. These products contained $10^4$ to $10^8$ cfu/g and were between 30 and 200 days.

Other products for which quantitative data were reported included mustard pickles and kimchi from Taiwan and pickled cucumbers from China, India, and the United States. Microbial counts ranged from undetectable (< $10^1$) to $10^8$ cfu/g. For several of these products, levels of species (e.g., Lactobacillus plantarum, Lactobacillus brevis, and Pediococcus cerevisiae) were reported. Species of Leuconostoc, Weissella and Lactobacillus were also reported for Korean kimchi, where they were generally present between $10^7$ and $10^8$ cfu/g.

# 9

# FERMENTED CEREALS

## 9.1 INTRODUCTION

Cereals are the edible seeds of the grass family of Poaceae, also known as Gramineae, and their cultivation dates to thousands of years ago. Wheat, maize, rice, barley, sorghum, millet, oat, and rye are the cereals most important on a global scale. Among them, wheat and rice represent the dominant crops, in Western and Asian countries, respectively. Cereals are one of the most important food sources for human consumption, with a production of more than 2 billion tons/year. However, unfortunately, roughly 30% of this amount is wasted or lost due to several reasons. Overall, food losses include all the edible parts discarded during the supply chain, while food wastes are residues of high organic load, removed during raw materials processing to foodstuff. In developing countries, substantial food losses occur during agricultural production, whereas in industrialized countries losses also include processed products during the distribution and consumption stages. Considering the unused food matrix as waste does not enforce the possibility of re-utilizing it in the food chain. For this reason, the use of the term "by-product" is increasing and identifies those wastes that become substrates for the recapture of functional compounds and the development of new products with a market value.

Fermented porridges and gruels are widely consumed in many African countries. Here, studies were reported from Burkina Faso, Uganda, Ghana, Benin, Tanzania, and Mexico (Table 1). These cereals were made using pearl millet, millet, sorghum, and maize as starting grains. In general, the cereals contained LAB and mesophilic aerobic bacteria with a range of $10^5$ to $10^9$ cfu/g.

In cereals processing, the two major by-products obtained during the traditional milling procedures are bran and germ. The initial purpose of milling was mostly to grind the grain, successively, the separation of the starchy endosperm from the outer layers (dry milling) became more important. The main reason for germ and bran removal is that, despite being rich in vitamins, minerals and dietary fiber, they adversely affect the processing properties of flours and this is one of the main reasons why the majority of cereal foods consumed are made of refined flour. However, refined flour is then lacking many compounds important for nutrition. Wet milling, on the other hand, is mainly used to produce starch and gluten, separating them from germ, bran and a precipitate solid fraction.

**Table 1: Fermented porridges and gruels are widely consumed in many African countries**

| Product (Region) | Source | Analyzed microorganisms | Count (log CFU/g) | Grain | CP | References |
|---|---|---|---|---|---|---|
| Ben-saalga (Burkina Faso) | Obtained from manufacturer | Total aerobic mesophiles | 7.1 | Pearl millet | 12 | Tou et al., 2006 |
| | | LAB | 7 | | | |
| | | Yeast | 5.5 | | | |
| Bushera (Uganda) | Markets | LAB | 8.1–8.4 | Millet | 5 | Muyanja et al., 2003 |
| | | LAB | 8.4 | Sorghum | 5 | |
| | | LAB | 8.9–9 | Millet and Sorghum | 5 | |
| Fura (Ghana) | Obtained from manufacturer | LAB | 6.6–8 | Does not specify | 8 | Owusu-Kwarteng et al., 2012 |
| Koko Sour Water (Ghana)[a] | Obtained from manufacturer | LAB | 8 | Does not specify | 3 | Lei and Jakobsen, 2004 |
| Mawè (Benin) | Market and manufacturer | Total aerobic mesophiles | 8.8 | Does not specify | 30 | Hounhouigan et al., 1993 |
| | | LAB | 8.9 | | | |
| | | Yeast | 6.4–6.9 | | | |
| Pozol (Mexico)[b] | Market | Total bacteria | 9.5 | Does not specify | 1 | Omar and Ampe, 2000 |
| | | LAB | 9 | | | |
| Togwa (Tanzania)[c] | Obtained from manufacturer | LAB | 9 | Sorghum, maize, millet, and maize | 36 | Mugula et al., 2003 |
| | | Yeast | 7 | | | |

[a] *Koko is porridge that have been heat treated. Koko sour water is the edible untreated water byproduct.*
[b] *Interpreted from graph. Measured outside and inside of sample in triplicate.*
[c] *Samples were obtained from manufacturer before fermentation.*

The employment cycle of these by-products can follow different paths (Fig. 1). The most common way of disposal is to use them as feed or for compost. However, to alleviate the environmental and economic burden of such losses, different approaches have been explored. One of them is biorefinery to produce biofuels such as ethanol. Ethanol is often produced from the cellulosic fractions of cereal bran, especially from maize. It is estimated that 4% of the global grains are used for ethanol production. Besides ethanol the production of chemical compounds such as lactic acid is also pursued. Lactic acid is traditionally applied in food industry as well as in pharmaceutical, textile and chemical industry. It can be obtained from an extensive range of carbon sources, of which cereal industry by-products such as maize cob, wheat bran, brewer's spent grains are very rich. Whereas, maize, rice and wheat bran are often used to produce phytic acid. After phytic acid extraction, the resulting by-product is deprived of an antinutritional factor, making it a more valuable ingredient in animal feed. Insoluble dietary fibers, fructans, antioxidants, and many other bioactive compounds are extracted from cereal by-products and used in food manufacturing. The oil obtained from wheat germ finds wide application in vitamin production, and cosmetic industry as well as in food, feed and as insect biological control agent, while the defatted wheat germ and wheat germ proteins are used as ingredients for several food products.

A more recent approach involves the generation, from rice and wheat bran, husk and straw, of nanoparticles exhibiting antibacterial activity or the production, through microbial fermentation, of biodegradable plastics. Cereal by-products are considered an ideal candidate to produce commercially important enzymes, due to their richness in nutrients but also to their low costs and wide availability as cultivation substrates. The use of cereal by-products is considered one of the most effective means in producing high value compounds.

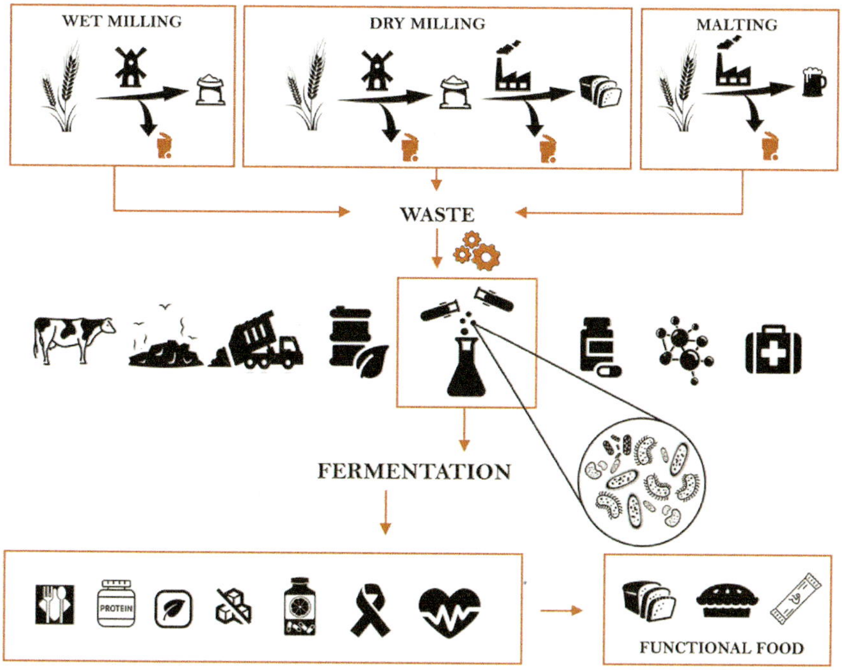


**Fig.1. The cereal industry by-products chain, from generation to the disposal, and their novel applications for improving nutritional and functional features of food.**

Besides enzymes production or bioplastics synthesis, microbial fermentation is a very efficient way to improve the nutritional and functional properties of cereal by-products to implement their use in food production. Fermentation, alone ore coupled with technological or biotechnological processing techniques, offers a variety of tools to modify cereal matrices. During fermentation, both endogenous and bacterial enzymes are able to modify the grain constituents affecting the structure, bioactivity, and nutrient bioavailability. Since fermentation with either lactic acid bacteria, yeasts or fungi is extensively employed to produce cereal-based foods with enhanced health properties its application to cereal by-products could, to a wider extent, improve the overall eco-sustainability of the food system, providing a suitable alternative to reduce malnutrition and hunger. In light of the above considerations, this review aims at providing a comprehensive overview of all the nutritional and functional positive outcomes deriving from fermentation of cereal industry waste and by-products.

## 9.2 FERMENTATION OF THE MILLING BY-PRODUCTS

### 9.2.1 Wheat Bran

The multiple outer layers of wheat (outer and inner pericarp, seed coat, and nucellar epidermis) are commonly referred to as bran. During conventional wheat roller milling, most of the endosperm is separated and further ground to wheat flour. Therefore, bran, together with the aleurone layer and remnants of endosperm, becomes a milling by-product. Different types of bran (i.e., coarse bran or regular bran, coarse weatings or fine bran, fine weatings or middlings or shorts, and low-grade flour or "red dog") can be distinguished depending on the particle size and the endosperm content. The most abundant polysaccharides of the bran layers, arabinoxylans and β-glucans, have a role in lowering the risk of type II diabetes and colorectal cancer as well as cardiovascular and diverticular diseases. However, bioactive compounds such as dietary fibers and phenolic acids are trapped in the cell wall structures resisting conventional milling and thus having low bioaccessibility. Thereby, new milling techniques, enzymatic treatments and fermentation processes targeting the structure of bran, have been studied with the aim of enhancing its nutritional potential.

Over the last years, the interest of the scientific community toward wheat bran fermentation, alone or combined with other approaches, markedly increased. Fermentation with two selected microbial strains combined with hydrolytic enzymes, mainly xylanase, endoglucanase, and β-glucanase was employed to obtain bran with higher nutritional quality than the native one. Bioprocessing influenced the microstructure of the bran, causing an extensive breakdown of the cell wall structures entailing an increase of the solubility of arabinoxylans, more than 11-fold compared to the native bran. These results confirmed those obtained in a previously study in which yeast fermentation of wheat bran led to a 66% increase in soluble arabinoxylans. Wheat bran was also fermented following a traditional fermentation process consisting of a back-slopping procedure of daily refreshments until a stable microbiota of lactic acid bacteria and yeasts was established. After fermentation, an increase up to 30% of soluble dietary fiber was obtained. Similar results were observed in a recent study fermenting bran with Lactobacillus bulgaricus and Streptococcus thermophilus combined with a commercial baker's yeast.

Wheat bran is also a rich source of proteins and amino acids, which, although different from those found in the endosperm, still have high biological and nutritional value. However, the bioavailability of proteins in bran is limited by multiple factors, (i) the structure of the layers (composed by insoluble and complexed carbohydrates and lignin) and (ii) the high content of antinutritional factors such as phytate which forms insoluble phytate-protein complexes. Because of lactic acid bacteria proteolytic activity and endogenous proteases activated by the low pH, increases of peptides and free amino acids concentration, including the functional non-protein γ-aminobutyrric amino acid (GABA), have been reported during fermentation. Consequentially, the in vitro digestibility of proteins, which gives information on their stability and on how they withstand to digestive processes, increased. The digestible protein fraction can also give information about the quality of the protein itself. It was indeed proven that the biological value, the ratio of essential amino acids, and the nutritional index were higher in fermented bran compared to the native one, and further increase of the above indexes was obtained when enzymes were used in combination with microbial fermentation.

The use of the starter cultures L. brevis E-95612 and Candida humilis E-96250, especially when cell-wall-degrading enzymes were added, affected not only the release and the composition of free amino acids, improving protein digestibility, but also enabled the release of phenols. Hydroxycinnamic acids are the most common phenolic acids found in wheat bran.

In particular, ferulic acid, a structural component of the cell walls of aleurone and pericarp is mainly esterified with arabinoxylans, and therefore has a very low bioaccessibility. The potential health effect of ferulic acid is ascribed to its antioxidant properties. In particular, its ability to inhibit the lipid peroxidation and the oxidation of low-density lipoprotein (the main cholesterol carrier in blood), is greater than other hydroxycinnamic acids. Ferulic acid also has anti-inflammatory effects. Bioprocessing of wheat bran with baker's yeast and an enzymes mixture containing ferulic acid esterase allowed the increase of free phenolic acids content. Spontaneous fermentation carried out by lactic acid bacteria (mainly belonging to the genera Lactobacillus, Leuconostoc and Pediococcus) and yeasts was also found to be effective on the release of ferulic acid, which increased of 82%. The release of ferulic acid occurred also during fungal fermentation. Among the edible mushrooms employed in a previous study, Hericium erinaceus was the one that allowed the highest release of ferulic acid (44% higher than the unfermented bran) due to the combined action of cellulase and ferulic acid esterase which were able to decompose wheat bran cell walls. An improvement in total phenol content and antioxidant activity toward free radical DPPH (2,2-diphenyl-1-picrylhydrazyl) and ABTS (2,20-azino-di-[3-ethylbenzthiazoline sulphonate]) occurred when wheat bran was fermented for 6 days at 30°C with Aspergillus oryzae MTCC 3107. Maximum antioxidant activity was noticed for fermented bran, in both methanol and ethanol extracts, compared to fermented wheat intact kernels and flour.

The role of oxygen during the fermentation of a liquid wheat bran sourdough was also investigated. The amount of oxygen influenced the microbial community, as well as the metabolite profile of fermented bran. It was indeed observed that anaerobic conditions, in which lactic acid bacteria and endogenous heterotrophic bacteria grew better, induced the conversion of ferulic and caffeic acids into their corresponding derivatives, and increased the amount of sinapic acid. On the contrary, wheat bran fermented in aerobic conditions, which favored yeasts growth, was characterized by the presence of the phenolic compounds dihydroxyphenyl ethanol and hydroxyphenylacetaldehyde. Moreover, a higher amino acids content was found after anaerobic fermentation compared to aerobic one.

Minerals and vitamins contained in wheat grains are mostly located in the bran fraction, especially in the aleurone layer. The bioavailability of minerals strongly depends on the content of phytic acid, which is generally very abundant in wheat bran and is considered an antinutritional factor. An improvement in phytase activity was observed when both fermentation and enzymes were used to bioprocess bran and a reduction of phytic acid was reported by several authors.

Several B-vitamins, mainly niacin, pantothenic acid, biotin, thiamin and small amount of riboflavin are present in wheat bran. Yeast fermentation successfully increased the folate content of wheat bran over 40%. In the study conditions, folate synthesis, which varies extensively between Saccharomyces cerevisiae strains, was also partially ascribed to the presence of indigenous lactic acid bacteria. The fortification in limiting vitamins was also proposed. For instance, the intake of vitamin B12, also known as cobalamin, is mainly possible through consumption of food from animal origin, therefore there is a risk of its deficiency for people consuming limited amount of animal food products. Thus, the fortification of plant-based food with this vitamin through fermentation represents a good strategy. Propionibacterium freudenreichii DSM 20271, one of the few microorganisms recognized as vitamin B12 producers, was used to ferment wheat bran. The content of the active form of vitamin B12 in fermented bran increased of about 5 times, and a higher content in riboflavin was also detected, proving that bran can be a potential substrate for vitamins synthesis.

The use of milling by-products in food processing entails technological drawbacks, which make their application more challenging. For instance, in wheat bread baking, gluten has a key role in the structure formation and bran addition weakens the gluten network structure, therefore affecting the gas-holding capacity of doughs. As a result, volume and elasticity of baked goods decrease.

An increase in the phenolic content and therefore in the antioxidant activity of bread containing wheat bran fermented with either yeasts or lactic acid bacteria was reported by few authors. Anson et al., fermented wheat bran with baker's yeast and used it to produce a fortified bread having almost 3-fold the amount of free ferulic acid and 8-fold when combined with hydrolytic enzymes treatment. p-coumaric and sinapic acids also increased with the processing. When subjected to gastro-intestinal digestion in vitro, despite the substantial increase in the bioavailability of phenolic compounds, mostly recovered from the jejunal compartment, only a small part of them was further metabolized in the colon section, especially those that were already partially degraded by the bran fermentation and enzymatic treatments. Colonic metabolites have been found to have anti-inflammatory properties suggesting that bread enriched in fermented wheat bran could show the same properties.

Fermented wheat bran was used to prepare a composite wheat-rye bread containing a β-glucan hydrogel. Although the flavonoid content was significantly higher in the experimental bread compared to the control, the concentration of phenolic acids decreased during sourdough fermentation. However, higher radical scavenging activity against both ABTS and DPPH radical were found. The same bread was found to lower the glucose response 120 min after the consumption in a small group of volunteers. Insulin response did not change compared to the control bread (wheat-rye bread), but the authors indicated that an increased amount of fermented wheat bran and β-glucan could further improve the nutritional impact of the wheat bran-rye bread.

## 9.2.2 Wheat Germ

Wheat germ is a high nutritional value by-product separated during the milling process. It is the primary source of vitamin E in wheat kernel and a rich source of vitamins of the group B, proteins, dietary fiber and minerals. Most of the essential amino acids are present in wheat germ proteins at concentrations higher than in the reference egg protein pattern. Wheat germ is also rich in unsaturated fatty acids, mainly oleic, linoleic and α-linoleic acids and functional phytochemicals especially flavonoids, sterols, octacosanols, and glutathione. However, its consumption is limited by some anti-nutritional factors (raffinose, phytic acid, and wheat germ agglutinin) and by the high lipase and lipoxygenase activity that favor lipid oxidation, negatively affecting the stability of wheat germ.

To solve this issue, the effects of sourdough fermentation on wheat germ stabilization were studied. Two lactic acid bacteria (Lactobacillus plantarum LB1 and Lactobacillus rossiae LB5) isolated from wheat germ were used as starters for sourdough fermentation. After 40 days of storage, compared to the raw germ, the fermented one had very low percentage of the aldehydes usually responsible for the rancidity perception, as well as of alcohols, ketones, furanones, and lactones, other volatile compounds occurring in lipid oxidation. The low pH achieved with fermentation was indeed responsible of the lower lipase activity. Fermentation also increased of ca. 50% the concentration of total free amino acids, more specifically Lys, the major limiting amino acid of wheat flour, and GABA were present in fermented wheat germ at the concentration of almost 2 g/kg. During sourdough fermentation of wheat germ, the phytase activity increased and an enhanced the bioaccesibility of $Ca^{++}$, $Fe^{++}$, $K^{+}$, $Mn^{++}$, $Na^{+}$, and $Zn^{++}$. Concomitantly, raffinose concentration decreased by 45% and a 33% increase in phenol content occurred, which resulted in higher scavenging activity toward free radical DPPH and ABTS. Antioxidant activity in food matrixes is often due to the presence of phenolic compounds; nonetheless, this functional property can also be ascribed to bioactive peptides. Biologically active peptides, often encrypted in the native sequence, can be produced from their protein precursor by digestive enzymes or during food processing.

The interest toward bioactive peptides from vegetable sources has increased thanks to the recent evidence of their wide potential functional effects. After 48 h of fermentation of a medium composed by 5% of defatted wheat germ, the maximum yield of peptides was obtained. The protein hydrolysate showed high antioxidant activity, determined as scavenging activity on DPPH, hydroxyl, and superoxide radicals.

One of the most promising features of fermented wheat germ is represented by the cytotoxic activity toward cancer cell lines. A commercially available wheat germ extract known as Avemar® is obtained by fermentation of wheat germ water-soluble extract with Saccharomyces cerevisiae, followed by concentration and drying. The anticancer properties of this extract have been shown in vitro on various human cancers cell lines (including leukemia, melanoma, breast, colon testicular, head and neck, cervical, ovarian, gastric, thyroid, and brain carcinomas), as well as on the prevention of chemical carcinogenesis, and some autoimmune conditions. These features are mainly attributed to two quinones, 2-methoxy benzoquinone and 2,6-dimethoxybenzoquinone which are naturally present in wheat germ as glycosylated and non-physiologically active form. For this purpose, selected strains of lactic acid bacteria possessing high β-glucosidase activity, therefore potentially able to release the two quinones, were used for wheat germ fermentation. During 24 h of incubation, the release of the non-glycosylated and physiologically active forms was almost complete. Compared to the control, the concentration of the above bioactive compounds increased up to 6-folds. While no effect was found for the raw wheat germ, the preparation fermented by Lb. plantarum LB1 and Lb. rossiae LB5 exerted anti-proliferative effect on human tumor cell lines (colon carcinoma and ovarian carcinoma), as showed by ex vivo assays.

Germ is rarely used for food processing mainly because of the short shelf-life, due to the presence of large amounts of unsaturated fat acids and of hydrolytic and oxidative enzymes, as well as to the negative effects on the technological properties of the flour. Fermented wheat germ, having lipase activity lower than raw wheat germ, was incorporated in bread effectively reducing the technological obstacles that prevent its use in baking. The concentration of total free amino acids, especially that of lysine, which is limiting in cereals, increased if compared to wheat bread. GABA content increased as well reaching 223 mg/kg. Among the nutritional benefits deriving from wheat germ fermentation and supplementation in bread manufacturing there were improved in vitro protein digestibility and decreased phytase activity, all without compromising baking properties. In a recent study, a sourdough composed by both wheat germ and bran was used to fortify wheat flour bread. In addition to the release of free amino acids and phenolic compounds, which reflected on the antioxidant activity, the fortified bread containing 15% of the above fermented milling by-products had 6.53% of dietary fiber, almost 5% higher than that of wheat flour bread. The in vitro and, especially, the in vivo glycaemic index were markedly lower in the fortified bread, reaching 36.9%, a value way below the threshold needed for a food product to be considered "low glycaemic index".

### 9.2.3 Rye Bran

Rye, one of the most important sources of dietary fiber in European Nordic countries, is often used as whole grain flour in the making of cereal based products. Nonetheless, rye bran is also a by-product of conventional milling and can be used as ingredient to increase food nutritional value. Besides fibers, the bran fraction is rich in many other bioactive compounds (phenolic acids, phytosterols, tocopherols), among which alkylresorcinols and steryl ferulates have been studied for their cancer preventive and antioxidant potential. The influence of fermentation conditions and type of bran (native or peeled) on the levels of bioactive compounds was studied. Bran fractions, deriving from native or peeled grains were fermented with baker's yeast for 6-20 h at temperatures ranging from 20 to 35°C. Fermentation of both peeled and native bran increased free ferulic acid and free total phenolics, indicating increased liberation of bound phenolic compounds from the polymeric rye bran structure. The increase in free phenolics after fermentation was reported to be 90% for native rye bran, and 30% for peeled bran. The level of folates increased over 100% with fermentation and the highest level was obtained when higher growth of indigenous lactic acid bacteria occurred, which was strongly dependent on fermentation conditions. However, the strong acidity was found to be deleterious for cinnamoyl esterases, the enzyme responsible for the release of ferulic acid esterified to the arabinofuranosyl residues. For this reason, the highest level of free ferulic acid was obtained in fermentation conditions with a pH value of 6–6.5.

Several authors studied the influence of rye bran bioprocessing on bread nutritional quality. When combining the action of hydrolytic enzymes and baker's yeast fermentation the main change was represented by the degradation of the cell wall structure which consequentially led to an increase in the solubility of dietary fibers, arabinoxylans, proteins, and free phenolics. As shown by the in vitro colon model, bioprocessing made carbohydrates more fermentable by the microbiota and allowed a higher release of ferulic acid. In a follow up study, the phenolic profile of bread fortified with bioprocessed rye bran was evaluated. p-cumaric, sinapic, and caffeic acids were detected at concentration 20–30-fold higher than the bread containing native rye bran. Additionally, although a slight alkylresorcinol degradation was observed, benzoxazinoid aglycones increased after the enzymatic bioprocessing. Despite the extensive phenolic acid release caused by the bioprocessing, when breads were subjected to in vitro colon model, only subtle differences were observed in the microbial metabolites.

Rye bran proved to be an excellent substrate for the synthesis of polyunsaturated fatty acids and β-carotene by fungal solid-state fermentation. Four Mucor spp. strains selected based on the ability to synthesize γ-linolenic acid (GLA) and β-carotene were used. Compared to oat flakes, barley groats and wheat bran, rye bran fermented with Mucor circinelloides CCF-2617 contained the highest content of both compounds, which was further increased by the addition of brewers' spent grain.

Rye and wheat bran were used as substrates for the synthesis of exopolysaccharides through lactic acid bacteria fermentation. Fermentation with Lactobacillus reuteri in the presence of sucrose enabled high glucan formation within 8 h of incubation of rye bran. In a recent study, two strains of Weissella confusa, known for the ability to produce significant amounts of dextran, were employed to ferment both wheat and rye bran. Rye bran proved to be an optimal substrate for in situ dextran production, reaching concentrations of 2–3% on dry matter. The in situ synthesis of EPS and oligosaccharides represents a way to obtain a food substrate with increased functional properties. In fact, they may act as prebiotic and have been shown to possess antitumor and immunomodulating activities in vitro. Additionally, EPS also act as hydrocolloids and can improve the technological properties of material like cereal bran, otherwise characterized by poor structure-forming capacity.

### 9.2.4 Rice Germ and Bran

Asian Countries are the major producers of rice, representing 50% of the daily energy supply of the diet of the local population. From the commercial white rice, germ and bran are removed because the oils they contain are quickly subjected to rancidity, reducing its shelf-life. It is estimated that every year 120,000 tons of rice husks alone are wasted worldwide. Rice milling by-products are currently underutilized, since their further exploitation is possible. Rice bran oils and proteins have demonstrated antioxidant properties and chronic disease preventing activity, particularly toward cardiovascular disease and certain cancers. However, the content of these bioactive compounds is not equally distributed among rice varieties. Microbial fermentation of rice by-products is an emerging area of scientific and industrial research. Rice bran fermented with S. cerevisiae was shown to exert anti-stress and anti-fatigue effects on rats. Moreover, water-soluble extracts of fermented rice bran had an anti-photoaging effect on human skin fibroblasts cultures. During the last decade, solid-state fungal fermentation of rice bran was extensively studied. The main results achieved concerned the increase of protein content and antioxidant activity, particularly efficient when the substrate had small particle size (0.18 mm). When defatted rice bran was fermented with Rhizopus sp. and Aspergillus oryzae, a high amino acids release and consequently a substantial (from 37.5 to 54.3%) chemical score increase were obtained. Apart from proteins, fibers and minerals, rice bran is a good source of oil, which can reach up to 20% of its weight. Fermentation with either Rhizopus oryzae or A. oryzae significantly increased palmitic and linoleic acids content, causing a decrease in saturated fatty acids and an increase in the unsaturated ones, thus improving the overall nutritional quality; additionally, when R. oryzae was used as starter for rice bran fermentation, a 10% reduction of total lipid content was also observed.

Several studies investigated rice bran anticancer properties. Rice bran fermentation with Lentinus edodes allowed the production of an exo-biopolymer, consisting of sugars (mainly arabinose, galactose, glucose, mannose, and xylose), uronic acid and a small amount of proteins. Oral administration of the polysaccharide extract induced the activation of natural killer cells and prolonged the life spans of mice transplanted with Sarcoma-180 cells, while inhibiting cancerous cells growth in the intraperitoneum. Metabolite formation from extracts of rice bran fermented with Saccharomyces boulardii were studied by Ryan et al.. It was indeed observed that fermentation altered bioactive compounds, reducing the growth of human B lymphomas in vitro. Another study showed that the extract obtained by co-fermentation of rice bran with Lactobacillus rhamnosus and S. cerevisiae, was able to reduce the cytotoxicity and inhibit melanogenesis in B16F1 melanoma through downregulation of microphthalmia-associated transcription factor. Anticancer properties were also found in brown rice bran fermented with Aspergillus oryzae. It was demonstrated that fermented rice bran acts as preventive agent against colon carcinogenesis in rats.

The antioxidant potential of fermented rice bran was also studied. Phenolic compounds are present in rice bran at high concentrations, however, 70% of them are esterified to the arabinoxylans present in the bran cell wall. Rice bran fermented by Issatchenkia orientalis, a yeast isolated from rice bran, showed higher free phenolic content compared to the native one. The extracts strongly inhibited reactive oxygen species generation and ameliorated oxidative stress-induced insulin resistance by neutralizing free radicals and upregulating adiponectin in adipocytes. Phenolic extracts obtained by solid-state fermentation of rice bran with Rhizopus oryzae, were evaluated for their ability to reduce free radical DPPH and inhibit the enzymes peroxidase and polyphenol oxidase. Compared to the native bran, the phenolic content doubled with fermentation and changed in composition, in fact, ferulic acid increased over 20-fold reaching 765 mg/g in fermented bran. Although no inhibition of the polyphenol oxidase enzyme was found, the phenolic extracts DPPH scavenging and peroxidase inhibitory activities. The release of phenolic compounds from the bran was also obtained by fermenting heat-stabilized defatted bran with Bacillus subtilis subsp. subtilis. Compared to the control, which only had low levels of p-cumaric and ferulic acids, fermentation allowed the release of gentistic, caffeic, syringic, p-coumaric, ferulic, sinapic, and benzoic acids. The potential of fungal solid-state fermentation on the release of bioactive compound having antioxidant activity was investigated. Changes in the phenolic profile by Rhizopus oligosporus and Monascus purpureus fermentation used as single or mixed starters were observed. Although total polyphenols content and ferric reducing ability of plasma increased upon fermentation, DPPH radical-scavenging activity decreased in some cases, due to a different composition in phenolic acids. Ferulic acid was the only phenolic acid present in all samples, before and after fermentation, whereas vanillic, caffeic, and 4-hydroxybenzoic acids were found only after fermentation. The use of the two fungi combined was the condition that allowed the highest release of ferulic acid.

The other by-product of rice milling is the germ, commonly separated by sieving and vibrating rice bran. The literature on rice germ is very limited, and studies on fermentation of rice germ initiated only few years ago. Extracts from rice germ fermented by the GABA-producing Lactobacillus sakei B2-16 accumulated 15% (of dry weight) of GABA and were found to have a positive impact on sleep disturbance in mice.

Regarding their use as food ingredients, the literature is very meager. Fermented rice bran was used in bread-making with the aim of balancing the lacking essential amino acids and enriching the protein content of wheat-based products. The authors evaluated the effect of different substitution levels of wheat with protein concentrates from natural and yeast fermented rice bran. At 10% substitution level, the composite bread had higher total amino acid content than control wheat bread. An increase in the radical scavenging activity and ferric reducing ability power were also observed.

## 9.3 MILLING BY-PRODUCTS FROM OTHER CEREALS

Barley and oat significantly differ in their chemical composition from other cereals; their cell walls are rich in the non-starchy polysaccharide β-glucan, which is the major component of the soluble dietary fiber, and has been associated with the reduction of plasma cholesterol and glycemic index, and a decreased risk of colon cancer. Despite the beneficial advantages deriving from the consumption of barley and oat dietary fibers, very little information in the literature deals with the fermentation of their by-products. Catechin and proanthocyanidins are among the polyphenol compounds contained in barley bran. Hordeumin, an anthocyanidin-tannin purple pigment produced from barley bran fermented using Salmonella typhimurium, was found to have antimutagen properties. Barley bran hydrolysates were used to obtain xylitol through bioconversion of xylose-containing solutions by the yeast Debaryomyces hansenii under microaerophilic conditions. Xylitol is employed in the food industry to manufacture sugar-free products because of its high sweetening power, anticaries properties, and its tolerance by diabetics.

Fermentation was used as means to enrich oat bran with folate. Folate is a generic name for several derivatives of pteroylglutamic acid (folic acid) and is necessary for methylation reactions in cell metabolism and for neural development of fetus during pregnancy. Oat bran was fermented with yeasts isolated from barley kernels and selected for the ability to synthesize folate, alone or together with lactic acid bacteria isolated from oat bran. The best folate producers were S. cerevisiae, followed by Pseudozyma sp., Rhodotorula glutinis, and Kluyveromyces marxianus. Many yeasts, beyond the considerable amount of folate produced, caused a decrease in the viscosity, suggesting a possible generation of soluble fibers, with positive repercussion on the nutritional effect. When inoculated together with Streptococcus thermophilus or L. rhamnosus, S. cerevisiae and Candida milleri produced significant amount of folates reaching 120 ng/g, suggesting that the consumption of 100 g of fermented oat bran could represents 15% of the recommended folates daily intake. Fermentation of oat bran with rye sourdough, previously obtained with a commercial starter culture containing lactobacilli and Candida milleri, allowed to double protein and β-glucan solubility. Since oat β-glucan can stimulate the growth of L. rhamnosus, a fermented oat bran suspension was used as a carrier for the probiotic strain. A simulator of the human intestinal microbial ecosystem (SHIME) was used to evaluate the effect on gut microbiota, concluding that Lb. rhamnosus colonized the SHIME and oat bran favored the growth of bifidobacteria.

Lactobacillus plantarum T6B10 and Weissella confusa BAN8 were used as selected starters to ferment maize milling by-products mixtures made with raw or heat-treated germ and bran. Lactic acid bacteria metabolisms improved the free amino acids and peptides concentrations as well as the antioxidant activity and induced phytic acid degradation. As previously reported for wheat germ, fermentation allowed the decrease of the endogenous lipase activity, stabilizing the matrix by preventing oxidative processes. When fermented maize by-products were used as ingredient for bread making (25% on total weight) dietary fiber and proteins content were of ca. 11% and 13% of dry matter, respectively. Compared to the use of the same amount of unfermented ones, the addition of pre-fermented maize by-products to bread caused a significant increase in protein digestibility (up to 60%), and a relevant decrease of the starch hydrolysis index (ca. 13%).

## 9.4 FERMENTATION OF THE CEREAL INDUSTRY WASTE

### 9.4.1 Brewers' Spent Grain

Brewers' spent grain (BSG) is recovered from mashing, one of the initial steps of brewing. During the boiling process, all soluble matter is extracted into the mash from the barley malt which, after lautering (or mash filtration), is separated into wort (liquid) and spent grain (solid) components. BSG represents 85% of the total residues from the brewing process and it is estimated that 30,000 tons of BSG are wasted yearly worldwide. The main current destination of BSG is cattle feed or discarded. BSG is rich in cellulose (17%) and non-cellulosic polysaccharides, especially arabinoxylans from the barley grain hull (39%). BSG contains up to 20% of proteins, particularly rich in lysine and approximately 30% of the total protein content is made of essential amino acids. Apart from uses in animal nutrition or recovery of valuable compounds such as carbohydrates, proteins and phenolic compounds, thanks to the health benefits associated with BSG ingestion, some attempts to exploit its use in food industry have also been made. The research so far has shown that due to BSG challenging technological properties, pre-treatments aiming at reducing its detrimental impact on food quality are necessary. Overall, milling and bioprocessing technologies, including fermentation, are potential means to use more of the BSG in food applications, conferring important health benefits. In one study, BSG was used as substrate to immobilize Lactobacillus casei suggesting that BSG can act as prebiotic, stimulating lactic acid bacteria growth.

A fermented liquid product from BSG having nutraceuticals properties was also developed. After a first substrate optimization phase aiming at improving bacterial growth and polyphenolic compounds release, L. plantarum ATCC 8014 was used to produce a fermented beverage displaying high antioxidant potential, due to the high content of total phenolic compounds and flavonoids released during the fermentation, as consequence of the acidification and microbial metabolism. Bread containing BSG fermented with a strain of L. plantarum, was positively judged for the main structural and sensory properties. Thanks to the high content of proteins, fibers and lysine, a 10% replacement of wheat flour with either BSG or fermented BSG, improved the nutritional properties of the resulting bread. Spontaneous fermentation of BSG also showed positive impact on bread properties, characterized by lower levels of phytic acid (more than 30% lower compared to the native counterpart) and higher antioxidant activity (up to 36%) compared to the bread containing unfermented BSG.

## 9.5 OTHER CEREAL-DERIVED WASTE: BREAD AND STARCH

On a global scale, one of the major food waste is bread, reaching thousands of tons daily. Industrial bread waste is generated at different stages: during the manufacturing process, because of substandard products or other processing factors such as crusts removal for sandwich bread production, or as unsold bread from retail. If not discarded or used as feed, over the last decades, wasted bread was used to produce chemicals, aroma compounds, enzymes and biofuels. The processes to produce a seasoning sauce from the hydrolysis of wheat bread or to produce a syrup from bioprocessed bread were even patented. The first attempt involving lactic acid bacteria fermentation was proposed a couple of decades ago. In this study, more than one hundred starters were screened based on the ability to acidify a medium containing bread crumbs. Three strains of L. plantarum, Staphylococcus carnosus, Pediococcus acidilactici and Micrococcus spp. were selected as able to resist for long fermentation periods (48–96 h) without producing off-flavors. A sourdough containing bread waste was produced with the aim of enhancing bread aroma and flavor. Fermentation with a commercial L. plantarum strain as starter (35°C for 48 h) of a dough made of 50% whole wheat bread crumb, favored the highest organic acids production. Although no specific nutritional features have been highlighted by the authors for the fermented bread, the role of the organic acids, the main responsible of reduced glycaemia and insulinemic responses, should be considered. Indeed, lactic acid lowers the rate of starch digestion in bread whereas acetic and propionic acids appear instead to prolong the gastric emptying rate. Therefore, it is assumable that fermented bread can have similar features of sourdough.

Starch production industry mostly interests cereals such as maize, rice, and wheat, and crops like potato, and like any other industry, has its own by-products. Broken rice, which is an inevitable by-product of rice milling, can be used to extract powder and crystal starch generating another by-product rich in proteins. The protein residue of starch extraction was treated with a combination of enzyme hydrolysis and microbial fermentation. The hydrolysate resulting from the action of proteolytic enzyme and a cell suspension of Bacillus pumilus AG1 displayed antioxidant activity toward ABTS radical. Almost all the peptides contained in the hydrolysate showed one or more features typical of well-known antioxidant peptides, most probably conferring a synergic antioxidant effect to the mixture with the potential to be used as functional ingredient for novel food formulations.

# 10

# EMULSIONS & GELS FERMENTATION

## 10.1 INTRODUCTION

All colloids possess an electrical charge which may be either positive or negative. In a given colloidal system, all particles have like charges and hence tend to repel one anotherand remain in suspension. To be insuspension state, the colloidal particles distribute themselves uniformly throughout the liquid in which they are dispersed.The colloidal state is a heterogeneous system consisting of two phase's. The particles of the colloidal substance constitute the dispersed phase while the liquid medium in which they are present isknown as dispersion medium.

Colloidal systems with two phases can occur in eight different combinations so colloidal systems are not restricted to the dispersion of solid in liquid only. Each of the three states of matter can be dispersed in a medium which may be gaseous, liquid or solid.

## 10.2 EMULSIONS

An emulsion is formed when two nonsoluble liquids (e.g., an oil and water) are agitated together to disperse one liquid into the other, in the form of drops. Emulsions can either be oil-in-water (O/W) or water-in-oil (W/O), depending on whether the continuous phase is the water or the oil, respectively. Drop sizes normally vary from 1 μm to 50 μm. When the agitation stops, if the drops coalesce and the two phases separate under gravity, the emulsion has been temporary. To form a stable emulsion, an emulsifying agent must be added to the system.

Sometimes, the formation of an emulsion is the deliberate outcome of a manufacturing process. This is the case, for example, in the production of mayonnaise, where ground mustard seeds are normally added to act as an emulsifying agent. Other times, the formation of an emulsion is totally undesirable. An example is the case of the oil industry where emulsification of oil and brine is common. It may occur in the oil reservoir itself or while flowing through pipelines, mechanical devices, such as pumps, and gas separators.

A mixture of two or more immiscible liquids. One liquid (the dispersed phase) is dispersed in the other (the continuous phase). Three substance arerequired to produce an emulsion: two immiscible liquids and an emulsifying agent. Prepared by shaking – oil/ water (milk), water/ oil (butter).

Emulsions are colloidal solutions in which both the dispersed phase and dispersion medium are liquids. However, the two liquids are immiscible, as miscible liquids will form true solution.

Emulsion are of two kinds:

1. **Oil-in-water emulsion:** Here the dispersed phase is oil while the dispersion medium is water. Milk is an example of this kind as in milk liquid fats are dispersed in water. Vanishing cream is another example.
2. **Water-in-oil emulsion:** Here dispersed phase is water and dispersion medium is oil. Butter, cod-liver oil, cold creams are examples of this type.

The liquids forming emulsion i.e. oil and water will separate out on keeping as they are immiscible. Therefore, an emulsifying agent or emulsifier is added to stabilize the emulsion. Soap is a common emulsifier. The preparation of emulsion in the presence of an emulsifier is called emulsification.

A substance that stabilizes an emulsion is called an emulsifier or emulgent. Emulsifiers work by increasing the kinetic stability of a mixture. Surfactants or surface active agents are one type of emulsifiers. Detergents are an example of a surfactant. Other examples of emulsifiers include lecithin, mustard, soy lecithin, sodium phosphates, diacetyl tartaric acid ester of monoglyceride (DATEM), and sodium stearoyl lactylate.

### 10.2.1 Distinction between Colloid and Emulsion

Sometimes the terms "colloid" and "emulsion" are used interchangeably, but the term emulsion applies when both phases of a mixture are liquids. The particles in a colloid can be any phase of matter. So, an emulsion is a type of colloid, but not all colloids are emulsions.

### 10.2.2 Examples of Emulsions

- Oil and water mixtures are emulsions when shaken together. The oil will form drops and disperse throughout the water.
- Egg yolk is an emulsion containing the emulsifying agent lecithin.
- Crema on espresso is an emulsion consisting of water and coffee oil.
- Butter is an emulsion of water in fat.
- Mayonnaise is an oil in water emulsion that is stabilized by the lecithin in egg yolk.

- The photosensitive side of photographic film is coated with an emulsion of silver halide in gelatin.

### 10.2.3 Preparation of Emulsions

Preparation of emulsions depends on the scale at which it is produced. On small scale mortar and pestle can be used but its efficiency is limited. To overcome these drawback small electric mixers can be used although care must be exercised to avoid excessive entrapment of air. For large scale production mechanical stirrers are used to provide controlled agitation and shearing stress to produce stable emulsions.

### 10.2.4 Dry Gum Method

In this method the oil is first triturated with gum with a little amount of water to form the primary emulsion. The trituration is continued till a characteristic 'clicking' sound is heard and a thick white cream is formed. Once the primary emulsion is formed, the remaining quantity of water is slowly added to form the final emulsion.

### 10.2.5 Wet Gum Method

As the name implies, in this method first gum and water are triturated together to form a mucilage. The required quantity of oil is then added gradually in small proportions with thorough trituration to form the primary emulsion. Once the primary emulsion has been formed remaining quantity of water is added to make the final emulsion.

### 10.2.6 Bottle Method

This method is employed for preparing emulsions containing volatile and other non-viscous oils. Both dry gum and wet gum methods can be employed for the preparation. As volatile oils have a low viscosity as compared to fixed oils, they require comparatively large quantity of gum for emulsification. In this method, oil or water is first shaken thoroughly and vigorously with the calculated amount if gum. Once this has emulsified completely, the second liquid (either oil or water) is then added all at once and the bottle is again shaken vigorously to form the primary emulsion. More of water is added in small portions with constant agitation after each addition to produce the final volume Proportions of.

**Table 1: Proportions of Oil, Water and Gum required for formation of primary emulsion**

| Type of Oil | Oil | Water | Gum |
|---|---|---|---|
| Fixed Oil | 4 | 2 | 1 |
| Mineral Oil | 3 | 2 | 1 |
| Volatile Oil | 2 | 2 | 1 |

## 10.3 ADVANTAGES AND DISADVANTAGES OF EMULSIONS AS DOSAGE FORMS

### 10.3.1 Advantages

- Unpalatable oils can be administered in palatable form.
- Unpalatable oil-soluble drugs can be administered in palatable form.
- The aqueous phase is easily flavoured.
- The oily sensation is easily removed.
- The rate of absorption is increased.
- It is possible to include two incompatible ingredients, one in each phase of the emulsion.

### 10.3.2 Disadvantages

- Preparation needs to be shaken well before use.
- A measuring device is needed for administration.
- A degree of technical accuracy is needed to measure a dose.
- Storage conditions may affect stability.
- Bulky, difficult to transport and prone to container breakages.
- Liable to microbial contamination which can lead to cracking.

### 10.3.3 Types of Emulsion

Emulsion is dispersion of one form of liquid into another form of liquid. Let's consider two types of emulsions:

**(1) Oil in water:** The emulsion in which oil is present as the dispersed phase and water as the dispersion medium (continuous phase) is called an oil-in-water emulsion. Milk is an example of the oil-in-water type of emulsion. In milk liquid fat globules are dispersed in water. Other examples are, vanishing cream etc.

Let's start with the discussion of oil dispersed in water. Here, dispersion medium is water and this kind of emulsion can be found in applications which have small quantity of fatty material like moisturizing creams. Now, let's discuss the case of milk which also lies in this type of emulsion. Milk consists of liquid fat which gets dispersed in water. These types of emulsions are unstable and hence can be separated into two different layers when at rest. Therefore, to balance the emulsion, another component called emulsifying agent is added. This emulsifying agent enables the medium and the suspended particles to mingle. These emulsifying agents can be gum, proteins, synthetic soaps, etc.

**(2) Water in oil:** The emulsion in which water forms the dispersed phase, and the oil acts as the dispersion medium is called a water-in-oil emulsion. These emulsion are also termed oil emulsions. Butter and cold cream are typical examples of this types of emulsions. Other examples are cod liver oil etc.

Now, let's discuss with water dispersed in oil. Here, dispersion medium is oil and its application can be found in butter and cream. Where do you find water in oil dispersion? We can find this in sunscreens as this type of emulsion has a higher sun protection factor and they are also better water resistant than oil in water emulsion.

Let's differentiate between the two items to give a clear view of the two types of emulsions. Consider butter and cream, where cream is oil in water emulsion and butter is water in oil emulsion. Literally, they are the same thing but butter is an emulsion resulting from the inversion of the cream which is an oil in water emulsion.

**Table 2: Difference between O/W and W/O emulsions**

| S.No | Oil in water emulsion (o/w) | Water in oil emulsion (w/o) |
|---|---|---|
| 1. | Water is the dispersion medium andoil is the dispersed phase | Oil is the dispersion medium and wateris the dispersed phase |
| 2. | They are non greasy and easilyremovable from the skin surface | They are greasy and not waterwashable |
| 3. | They are used externally to providecooling effect e.g. vanishing cream | They are used externally to preventevaporation of moisture from thesurface of skin e.g. Cold cream |
| 4. | Water soluble drugs are more quicklyreleased from o/w emulsions | Oil soluble drugs are more quicklyreleased from w/o emulsions |
| 5. | They are preferred for formulationsmeant for internal use as bitter taste ofoils can be masked. | They are preferred for formulationsmeant for external use like creams. |
| 6. | O/W emulsions give a positiveconductivity test as water is theexternal phase which is a goodconductor of electricity. | W/O emulsions go not give a positiveconductivity test as oil is the externalphase which is a poor conductor ofelectricity. |

## 10.4 CLASSIFICATION OF EMULSIFYING AGENTS

Emulsifying agents can be classified as:

### 10.4.1 Natural Emulsifying Agents

A large number of emulsifiers are natural products derived from plant or animal tissue. Most of the emulsifiers form hydrated lyophilic colloids (called hydrocolloids) that form multimolecular layers around emulsion droplets. Hydrocolloid type emulsifiers have little or no effect on interfacial tension, but exert a protective colloid effect, reducing the potential for coalescence, by:

- Providing a protective sheath around the droplets
- Imparting a charge to the dispersed droplets (so that they repel each other)
- Swelling to increase the viscosity of the system (so that droplets are less likely to merge)

Natural emulsifying agents from vegetable sources: These consist of agents which are carbohydrates and include gums and mucilaginous substances. Since these substances are of variable chemical composition, these exhibit considerable variation in emulsifying properties. They are anionic in nature and produce o/w emulsions. They act as primary emulsifying agents as well as secondary emulsifying agents (emulsion stabilizers). Since carbohydrates acts a good medium for the growth of microorganism, therefore emulsions prepared using these emulsifying agents have to be suitable preserved in order to prevent microbial contamination. E.g. tragacanth, acacia, agar, chondrus (Irish Moss), pectin and starch.

### 10.4.2 Gum Acacia

It is generally used in the concentration of 8-15% and gives a stable and palatable emulsion over the pH range of 2-10. Emulsions tend to cream using this as the viscosity is low.

### 10.4.3 Tragacanth

Used in the concentration range of 1-2%. Rarely used now because it forms thick and coarse emulsions.

### 10.4.4 Agar

Used in the concentration of 2%.

### 10.4.5 Starch

Used in concentration of 2-5%. Rarely used because it forms coarse emulsions.

### 10.4.6 Pectin

Used as 1%. Acts as emulsion stabiliserin acacia emulsion. Natural emulsifying agents from animal source: The examples include gelatin, egg yolk and wool fat (anhydrous lanolin). Type A gelatin (Cationic) is generally used for preparing o/w emulsion while type B gelatin is used for o/w emulsions of pH 8 and above. Lecithin and cholesterol present in egg yolk also act as emulsifying agent. They show surface activity and are used for formulating o/w emulsions. However they are used only for extemporaneous preparation and not for commercial preparation as it darken and degrade rapidly in unpreserved systems. Wool fat is mainly used in w/o emulsions meant for external use. They absorb large quantities of water and form stable w/o emulsions with other oils and fats. Animal derivatives are more likely to cause allergic reactions and are subject to microbial growth and rancidity. Their advantage is their ability to support formation of w/o creams.

### 10.4.7 Gelatin

Used in concentration of 1%. Two grades are available-Pharmagol A which has an acidic pH and Pharmagol B, which has an alkaline pH. Emulsions prepared have agreeable taste but are prone to bacterial contamination. Gelatin is used for preparing o/w emulsions.

### 10.4.8 Egg Yolk

It contains lecithin which is a phospholipids acting as an emulsifier. Used in concentration range of 12-15%. Mainly used for extemporaneous preparation as long term stability is a problem. It forms stable o/w emulsions, which needs to be refrigerated to prolong shelf life.

### 10.4.9 Methyl Cellulose

This is non ionic in nature and is stable over a wide pH range. It is mainly used for emulsification of mineral and vegetables oil. Drawback is that it gets precipitated in the presence of large quantities of electrolytes.

### 10.4.10 Alkali Soaps

Produce good oil in water emulsions. Unstable at pH below 10 and are incompatible with acids and polyvalent inorganic and long chain organic cations.

### 10.4.11 Amine Soaps

Used for preparing o/w emulsions. Stable to the presence of calcium ions or changes in pH

### 10.4.12 Bold Text Metallic Soaps

Generally used for formulating w/o emulsions. They are usually insoluble in water

### 10.4.13 Alkyl Sulphates and Phosphates

They form o/w emulsions.

### 10.4.14 Alkyl Sulponates

Used mainly as wetting agents and form stable o/w emulsions along with a secondary emulsifying agent.

Cationic surfactants: The positive charge cations produced on dissociation are responsible for emulsifying properties. They are mainly used in external preparations such as lotions and creams. Quaternary ammonium compounds such as cetrimide, benzalkonium chloride and benzethonium chloride are examples of important cationic surfactants. These compounds besides having good antibacterial activity are also used in combination with secondary emulsifying agents to produce o/w emulsions for external application.

Non-ionic surfactants: They are the class of surfactants widely used as emulsifying agents. They are extensively used to produce both oil in water and water in oil emulsions for internal as well as external use. The emulsions prepared using these surfactants remain stable over a wide range of pH changes and are not affected by the addition of acids and electrolytes. They also show low irritancy as compared to other surfactants. E.g. glyceryl esters such as glyceryl monostearate, propylene glycol monostearate, macrogol esters such as polyoxyl stearates and polyoxyl-castor oil derivatives, sorbitan fatty acid esters such as spans and their polyoxyethylene derivatives such as tweens (polysorbates).

Ampholytic surfactants: These are the substances whose ionic charge depends on the pH of the system. Below a certain pH, these are cationic while above a defined pH, these are cationic. At intermediate pH these behave as zwitterions. e.g. lecithin.

**Finely Divided Solids:** This group consists of finely divided solids having balanced hydrophilic lipophillic properties. They accumulate at the oil/water interface and form a coherent interfacial film around the droplets of dispersed phase globules and prevent coalescence. If the solid particles are preferentially wetted by oil, a w/o emulsion is formed while if wetting is done by water then o/w emulsion is seen. e.g., bentonite, aluminium magnesium stearate, attapulgite, colloidal anhydrous silica and hectorite. The emulsions formed using finely divided solids are stable and less prone to microbial contamination.

**Auxillary Emulsifying Agents:** A variety of fatty acids (e.g., stearic acid), fatty alcohols (e.g., stearyl or cetyl alcohol), and fatty esters (e.g., glyceryl monostearate) serve to stabilize creams through their ability to thicken the emulsion. Because these agents have only weak emulsifying properties, they are always used in combination with other emulsifiers.

## 10.5 PROPERTIES OF EMULSIONS

Emulsions usually appear cloudy or white because light is scattered off the phase interphases between the components in the mixture. If all of the light is scattered equally, the emulsion will appear white. Dilute emulsions may appear slightly blue because low wavelength light is scattered more. This is called the Tyndall effect. It's commonly seen in skim milk. If the particle size of the droplets is less than 100 nm (a microemulsion or nanoemulsion), it's possible for the mixture to be translucent.

Because emulsions are liquids, they don't have a static internal structure. Droplets are distributed more or less evenly throughout a liquid matrix called the dispersion medium. Two liquids can form different types of emulsions. For example, oil and water can form an oil in water emulsion, where the oil droplets are dispersed in water, or they can form a water in oil emulsion, with water dispersed in oil. Further, they can form multiple emulsions, such as water in oil in water.

Most emulsions are unstable, with components that won't mix on their own or remain suspended indefinitely.

### 10.5.1 Structure of Emulsions

When a liquid e.g. oil is dispersed in another immiscible liquid e.g. water (continuous phase) by mechanical agitation, an emulsion is obtained. Afood emulsion basically is a two – phase system. Most of the dispersed droplets in foodemulsions have diameters between 0.1 and 10 micrometer. Emulsion stability can be increased by adding an emulsifying agent. An emulsifyingagent is a surface active agent that lowers the interfacial tension and forms physical barriers around each droplet so as to prevent their coalescence (combination of smalldroplets to form large droplets).

## 10.6 APPLICATIONS OF EMULSIONS

1. Emulsions play very important role in our daily life.
2. The cleansing action of soap and synthetic detergents for washing clothes, bathing, etc is based upon the formation of oil in water type emulsion.
3. Milk is an emulsion of fat in water. Milk cream and butter are also emulsions.
4. Various cold creams, vanishing creams, body lotions, etc. are all emulsions.
5. Various oily drugs such as cod liver oil are administered in the form of emulsion for their better and faster absorption. Some ointments are also in the form of emulsions.
6. The digestion of fats in the intestine occurs by the process of emulsification.
7. Emulsions are used for concentrating the sulphide ores by froth flotation process.

Finely powdered ore is treated with an oil emulsion and the mixture is vigorously agitated by compressed air when the ore particles are carried to the surface and removed.

### 10.6.1 How Emulsification Works?

There are a few mechanisms that may be involved in emulsification:

- Emulsification may occur when the interfacial surface tension between two liquids is reduced. This is how surfactants work.
- An emulsifier may form a film over one phase in a mixture to form globules that repel each other, allowing them to remain evenly dispersed or suspended.
- Certain emulgents increase the viscosity of the medium, making it easier for the globules to remain suspended. Examples include the hydrocolloids acacia and tragacanth, glycerine, and the polymer carboxymethyl cellulose.

## 10.7 FORMATION AND STABILITY OF EMULSIONS

Controlling factors in the formation of an emulsion are: mechanical energy, agitation time, temperature, volumetric ratio between the two phases, degree of dispersion of the internal phase and presence of impurities or surfactants. The material of the shearing plates for the homogenizer used in the emulsification process also influences the type of emulsion formed, e.g., oil-wetted plates strongly favor W/O emulsions.

There are many ways of producing an emulsion and it is usually achieved by applying mechanical energy through agitation, normally by using an homogenizer. Initially, the interface between the two phases is deformed and large droplets are formed. These droplets are subsequently broken up into smaller droplets by the continuing agitation. Impurities or surfactants present in the system adsorb at the interfaces of the droplets, lower the interfacial tension, and thereby facilitate coalescence. However, the surfactant film formed at the interface of the droplet also tends to resist coalescence. A detailed review of the principles of emulsion formation has been published recently by P. Walstra (1993), where droplet break-up in laminar and turbulent flow is discussed and quantitative relations are presented.

The stability of an emulsion is dependent on the magnitudes of the previously-mentioned opposing effects and is affected by: interfacial viscosity, electric charge on drops, droplet size and concentration, and viscosity of the continuous phase. Aging of an emulsion may also affect its stability as the nature of the interfacial film, which helps to keep it stable, can change with time.

**Mechanical Force:** To make an emulsion you first need to apply a mechanical force to break down the dispersed phase into small droplets that become suspended in the continuous phase. For this you can use a whisk, blender or other lab equipment such as rotor-stator homogenizer. We'll cover these later.

**Emulsifier:** The next problem to solve is to make the emulsion stable. The higher the force you applied when making the emulsion, the smaller the droplets and the more stable the emulsion is. However, no matter how small the droplets are, the ingredients will eventually separate without the presence of an emulsifier that keeps the molecules with different polarity from repelling each other.

**Thickener:** Finally, adding a thickener to the continuous phase can make the emulsion even more stable as this makes it more difficult for the dispersed droplets to move and combine.

Any mixture of oil and water can be transformed into an emulsion, even without emulsifiers. However, without emulsifiers, the mixture will also quickly separate back into its immiscible (not mixable) parts. You can try a simple experiment with oil and water. Quickly whisk together oil and water, and it will eventually turn into an cloudy mixture. You'll notice plenty of bubbles, however, and those bubbles will quickly grow larger until the water and oil are completely separate.

If, however, you use the emulsifier mustard and thickening agents like honey or xanthan gum, it's feasible to mix a stable vinaigrette by hand. The dressing may separate over time, but will likely stay emulsified through a dinner service.

## 10.8 PRESERVATION OF EMULSIONS

### 10.8.1 Preservation from Micro-Organisms

It is necessary to preserve the emulsions from microorganisms as these can proliferate easily in emulsified systems with high water content, particularly if carbohydrates, proteins or steroidal materials are also present.

Contamination due to microorganisms can result in problems such as color and odor change, gas production, hydrolysis, pH change and eventually breaking of emulsion. Therefore is necessary that emulsified systems be adequately preserved. An ideal preservative should be nonirritant, nonsensitizing and nontoxic in the concentration used. It should be physically as well as chemically compatible with other ingredients of the emulsions and with the proposed container of the product. It should not impart any taste, color or odor to the product. It should be stable and effective over a wide range of pH and temperature. It should have have a wide spectrum of activity against a range of bacteria, yeasts and moulds. The selective preservative should have high water solubility and a low oil/water partition coefficient. It should have bactericidal rather than bacteriostatic activity.

Examples of antimicrobial preservatives used to preserve emulsified systems include parahydroxybenzoate esters such as methyl, propyl and butyl parabens, organic acids such as ascorbic acid and benzoic acid, organic mercurials such as phenylmercuric acetate and phenylmercuric nitrate, quarternary ammonium compounds such as cetrimide, cresol derivatives such as chlorocresol and miscellaneous agents such as sodium benzoate, chloroform and phenoxyethanol.

### 10.8.2 Preservation from Oxidation

Oxidative changes such as rancidity and spoilage due to atmospheric oxygen and effects of enzymes produced by micro-organisms is seen in many emulsions containing vegetables and mineral oils and animal fats. Antioxidants can be used to prevent the changes occurring due to atmospheric oxygen.

Antioxidants are agents having a high affinity for oxygen and compete for it with labile substances in the formulation. The ideal antioxidant should be nontoxic, nonirritant, effective at low concentration under the expected conditions of storage and use, soluble in the medium and stable. Antioxidants for use in oral preparation should also be odorless and tasteless.

Some of the commonly used antixidants for emulsified systems include alkyl gallate such as ethyl, propyl or dodecyl gallate, butylated sshydroxyanisole (BHT), butylatedhydroxytoluene (BHT)

### 10.8.3 Choice of Emulsifiers

The choice of surfactants for a particular process depends on the restrictions established for that particular application, e.g., in the food industry, emulsifying agents must be edible. Another factor to be considered in the choice of a stabilizing agent is whether the desired type of emulsion is an O/W or W/O, as the stabilizing agent largely determines which phase is the continuous one. The phase in which the surfactant is more soluble will become the continuous phase.

Emulsions are often used in the most diverse fields, e.g., food industry, pharmaceutical products and manufacture of lubricants.

Some examples of emulsifiers are soaps, proteins, starch and gelatine.

### 10.8.4 Separation of Emulsions

Methods normally used to break emulsions are:

- **Gravity settling:** Settling of emulsions is more rapid when the drop size is larger and when the continuous phase viscosity is lower. For faster separation, heat can be applied to reduce the viscosity of the continuous phase and sometimes to reduce the effectiveness of the surfactant.
- **Centrifugation:** Faster separation by increasing the centripetal acceleration force.
- **Electrical coalescence:** The application of an electrical current (direct or alternating) causes the internal phase droplets to coalesce.
- **Chemical methods:** Coalescence can be achieved by the addition of suitable chemicals. For instance, by adding electrolytes the charge at the droplets' interfaces may be neutralized and coalescence can result.

A combination of the above methods may also be chosen: heat to modify the continuous phase, chemistry to modify the emulsion, and electricity to finalize the separation.

## 10.9 QUALITY CONTROL TESTS FOR EMULSIONS

The following are the quality control tests done for emulsions:

**1. Determination of particle size and particle count:** Determination of changes in the average particle size or the size distribution of droplets is an important parameter used for the evaluation of emulsions. It is performed by optical microscopy, sedimentation by using Andreasen apparatus and Coulter counter apparatus.

**2. Determination of viscosity:** Determination of viscosity is done to assess the changes that might take place during aging. Emulsions exhibit non-newtonian type of flow characterstics. The viscometers which should be used include cone and plate viscometers. Capillary and falling sphere type of viscometrs should be avoided. For viscous emulsions, the use of penetrometer is recommended as it helps in the determination of viscosity with age. In case of o/w emulsions, flocculation of globules causes an immediate increase in viscosity. After this change, the consistency of the emulsion changes with time. In case of w/o emulsions,, the dispersed phase particles flocculate quite rapidly resulting in a decrease in viscosity, which stabilizes after 5 to 15 days. As a rule, a decrease in viscosity with age reflects an increase of particle size due to coalescence.

**3. Determination of phase separation:** This is another parameter used for assessing the stability of the formulation. Phase separation may be observed visually or by measuring the volume of the separated phases.

**4. Determination of electrophoretic properties:** Determination of electrophoretic properties like zeta potential is useful for assessing flocculation since electrical charges on particles influence the rate of flocculation. O/W emulsion having a fine particle size will exhibit low resistance but if the particle size increase, then it indicates a sign of oil droplet aggregation and instability.

### 10.9.1 Stability Testing

Stability of emulsions is an important parameter for the formulator. Stability testing of emulsions involves determining stability at long term storage conditions, accelerated storage conditions, freezing and thawing conditions. Stress conditions are applied in order to speed up the stability testing. The stress conditions used for speeding up instability of emulsions include:

- Centrifugal force
- Agitational force
- Aging and temperature

The following physical parameters are evaluated to assess the effect of any of the above stress conditions:

- Phase separation
- Viscosity
- Electrophoretic properties
- Particle size and particle count

**Particle size and size distribution:** The freeze-thaw cycling technique used to assess emulsions for stress testing for stability testing result in increase of particle growth and may indicate future state after long storage. It is of importance to study the changes for absolute particle size and particle size distribution. It is performed by optical microscopy, sedimentation by using Andreasen apparatus and Coulter counter apparatus. None of these methods are direct methods. However microscopic method allows the observer to view the actual particles.

**Rheological studies:** Cone and Plate viscometer with variable shear stress control can be used for evaluating viscosity of emulsions.

## 10.10 GELS FERMENTATION

A gel is formed when the particles cross-link, using chemical bonds, into a three dimensional network that behaves like a solid, even though it is mostly liquid. Cooked egg white is a gel formed when the proteins in the egg white open up from their natural folded shape, and then bind to one another in many places.

A gel is a sol in which the solid particles are meshed such that a rigid or semi-rigid mixture results.

Gels are the type of colloids in which the dispersed phase is a liquid and the dispersion medium is a solid. Cheese, jelly, boot polish are common examples of gel. Most of the commonly used gels are hydrophilic colloidal solution in which a dilute solution, under suitable conditions set as elastic semi solid masses. For example, 5% aqueous solution of gelatin in water on cooling forms the jelly block.

Gels may shrink on keeping by loosing some of the liquid held by them. This is known as syneresis or resetting on standing. Gels are divided in two categories: elastic gels and non elastic gels. Elastic gels are reversible. When partly dehydrated on loosing water, they change back into the original form on addition of water. The non elastic gels are not reversible. Gels are useful in many ways. Silica, cheese, jelly, boot polish, curd are commonly used gels. Solidified alcohol fuel is a gel of alcohol in calcium acetate.

Examples: Fruit jelly is an example of a gel. Cooked and cooled gelatin is another example of a gel. The protein molecules of gelatin crosslink to form a soid mesh which contains pockets of liquid.

### 10.10.1 Hydrogels

Because hydrogels have a degree of flexibility similar to natural tissue, they are often used as scaffolds in tissue engineering and may contain human cells in order to repair tissue. Hydrogels consist of more than 99 percent of water and are superabsorbent. The practical uses of hydrogels include breast implants, granules for holding soil moisture in arid areas and dressings for burns and hard-to-heal wounds.

### 10.10.2 Organogels

Organogels or organic gelulators are noncrystalline, nonglassy, thermoplastic solid materials composed of a liquid organic phase entrapped in a three-dimensionally, cross-linked network. Organogels are frequently found in food processing, pharmaceuticals and cosmetics.

### 10.10.3 Xerogels

Xerogels are solids formed from gels that do not suffer from shrinkage in the process. Xerogels generally retain a high degree of porosity. Rubber and gelatin are examples of xerogels.

### 10.10.4 Aerogels

Aerogels are low-density materials created when the liquid component of a gel has been replaced by nitrogen or air. This results in very low density solids that are highly effective as thermal insulators. Aerogels are used as vehicles for drug prophylaxis.

Gels are classified into elastic and non elastic gels:

**Elastic gels:** These are those gels which posses the property of elasticity i.e. they change to solid mass on dehydration which can again be converted into gel by addition of water followed by heating and cooling. When they are placed in contact with water they absorb water and swell. This property is known as imbibitions. Examples of Elastic gels are gelatin, agar, starch etc.

**Non-elastic gels:** These are those gels which do not possess the property of elasticity i.e. they change to solid mass on dehydration which becomes rigid and cannot be converted into the original form by heating with water. They do not show the phenomenon of imbibitions. Best example of these gels is silicic acid.

## 10.11 PROPERTIES OF GELS

Many gels display thixotropy – they become fluid when agitated, but resolidify when resting. In general, gels are apparently solid, jelly-like materials. By replacing the liquid with gas it is possible to prepare aerogels, materials with exceptional properties including very low density, high specific surface areas, and excellent thermal insulation properties.

### 10.11.1 Gel Aesthetic Properties

Aesthetic properties are the properties of gelling agents that determine how the end consumer (the diner eating the food) perceives them.

### 10.11.2 Gel Mouthfeel, Brittleness, Elasticity and Stiffness

Mouthfeel is the broad, difficult to define category that people use to describe a wide range of sensations. With thickeners, we talked about viscosity and creaminess, and while these do matter to gels, we'll talk more about brittleness, elasticity, and stiffness.

In edible gels, brittleness and elasticity are usually related. Elasticity is defined by the ability of a material to return to its original shape after it is deformed. Consider the difference between Play-Doh (not elastic) and a water balloon (elastic). Brittleness is defined by the amount an object will deform before it breaks. Think about the difference between a new rubber band and one that has been dried out for a few months. The dry rubber band snaps after much less pulling than the new one.

While there are inelastic materials that have a brittleness independent of their elasticity (like metals, for example), edible gels are almost all elastic, so you can think of elasticity and brittleness as opposites when dealing with edible gels.

Be sure not to confuse elasticity and brittleness with stiffness. Although these qualities are perceived in similar ways by the diner, chefs should understand the distinction. Whereas elasticity and brittleness are innate properties of a hydrocolloid, stiffness is defined by the concentration of hydrocolloid to water. For example, a cube of gelatin is made with the same ingredient (gelatin) as a gummy candy, but the two have very different textures that depend on the how much gelatin is used.

### 10.11.3 Gel Appearance Clarity

And let's not forget the final aesthetic quality appearance. Some gels are more transparent than others. This property is known as clarity. For example, gelatin forms a clear gel while agar forms a semi-transparent gel.

### 10.11.4 Gel Functional Properties

The whole point of gelling agents is to create textures from ingredients that are normally liquids. In order for chefs to create novel textures, they have to understand the capabilities and limitations of the gelling agents they are using.

### 10.11.5 Dispersion and Hydration Temperature

The most important functional concern is the temperature at which hydrocolloids disperse and hydrate. In the cornstarch example, note that cornstarch will disperse in cold water but does not hydrate until heated. If, on the flip side, you tried to disperse corn starch in hot water, some of the cornstarch would rapidly hydrate and form protective shells around powdered cornstarch within, resulting in lumps of the thickener in your liquid. In addition, cornstarch will adopt very different textures depending on the temperature you choose to serve it at. We extensively discuss the temperature properties of many hydrocolloids in our hydrocolloid guide.

### 10.11.6 Gelling and Melting Temperature

Gelling agents have a few special temperature considerations. All gels have a temperature at which they will gel and melt. Let's look at the example of plain gelatin.

Gelatin should be dispersed in room-temperature or cold water to prevent lumps. It hydrates at temperatures near the boiling point of water. Once the gelatin liquid has been heated, it forms a gel at refrigerator temperatures. The gel begins to melt at the upper range of reasonable room temperature.

And some strange things can happen at other temperatures as well as we explain below:

**Gel syneresis and freeze-thaw stable gels:** If you were to freeze gelatin and then place it in the refrigerator, much of the liquid held in the gel would leak out. This is called "syneresis" and it's generally considered an undesirable quality. Not all gels exhibit syneresis after freezing, but usually those gels must be stabilized with a combination of hydrocolloids. A gel that does not exhibit syneresis after freezing is called "freeze-thaw stable."

## 10.12 THERMOREVERSIBLE AND THERMOIRREVERSIBLE GELS

Freeze-thaw stability should not be confused with thermoirreversibility. If a gel is thermoreversible, it can be converted back into a liquid form through a simple temperature change. Gelatin, for example, is thermoreversible. Egg white, on the other hand is thermoirreversible. Once an egg white has set, there is no way to change it back into liquid egg whites using temperature alone.

### 10.12.1 Gel Hysteresis

Some gels demonstrate a property known as hysteresis. Explained simply, hysteresis is a range of temperatures in which a gel is likely to stay in its current state. When we think of solids and liquids, we think of a static temperature at which a solid turns into a liquid and vice versa. For example, water turns into ice at 32°F/0°C and ice turns into water at the same temperature. But, some gels gel at a temperature different from the temperature at which they melt. For example, agar agar melts at 185°F/85°C and forms a gel at 88°F/32°C. The state of agar in between this range depends on the state that it is currently in.

### 10.12.2 Fluid Gels

When we talk about using blenders with hydrocolloids, we usually mean using these machines to help hydrate a hydrocolloid. But some hydrocolloids display a unique characteristic enabled through a blender. If you were to blend a gel made of gelatin or carrageenan, the force of the blades (called shear) would turn the gellified gelatin into a liquid. If the temperatures were right, that liquid would slowly convert back into a gel. This shear thinning property is called thixotropy. Under static conditions, these hydrocolloids produce solid gels (or thick viscous fluids) that will start flowing and become less viscous when agitated, stirred, shaken or blended. If let to rest, these fluids would return to their original solid state or thick viscosity. The time it takes for this transformation depends on the hydrocolloid. It could be instant or it could take a few hours.

Other hydrocolloids such as agar agar, become fluid when blended but do not return back to their original solid gel state. When blending an agar agar gel, it breaks down into tiny pieces but will not revert back into a solid. We refer to this by saying that agar agar forms a fluid gel under shear. A fluid gel is still a gel, but because of the smaller particle size, it behaves in many ways like a thickened fluid. Agar agar and gellan gum are the most common hydrocolloids used to make fluid gels.

Fluid gels have both the properties of solids and liquids. Gels are usually solidified liquids that hold its shape thanks to their molecular structure. On the other hand, liquids are fluid and do not hold their shape given their weak molecular structure. But fluid gels can hold their shape while they are also fluid and can be spread or reshaped into various forms.

### 10.12.3 Gel Flavor Release

Some gels have better flavor release than others. In general, the softer the gel the better the flavor release. High-fat gels and starches tend to have a slow and long-lasting flavor release while gelatins, pectins and gellan gels tend to have a fast and short-lived flavor release.

### 10.12.4 Hydrocolloid Interactions

The final functional properties all deal with how a hydrocolloid behaves in the presence of other things. For example, will a gel form when used in conjunction with sugar, alcohol, and acid?

Thickeners can also demonstrate synergies with other hydrocolloids as well as in the presence of ions. For example, neither locust bean gum nor xanthan gum form gels alone, but will form a gel when you combine the two hydrocolloids. Another example: carraggeenan is stronger in the presence of calcium, which is why it is often used in calcium-rich dairy applications.

### 10.12.5 Stability

Dry gelatin has an almost infinite shelf life as long as the moisture content is such as to ensure that the product is stored below the glass transition temperature.

The stability of gelatin in solution depends on temperature and pH. Generally, to minimise loss of gel strength and viscosity with time, the pH of the solution should be in the range 5 to 7 and the temperature should be kept as low as possible, consistent with the avoidance of gelation and the suitability of the solution viscosity to the particular application. Often the cause of degradation or hydrolysis of gelatin in solution is microbial proliferation, so gelatin solutions should not be stored for longer than is absolutely necessary, and after addition of the acid to confectionery formulations, the solution should be used and cooled/gelled with minimal delays.

### 10.12.6 Microencapsulation - Mixed Film Forming Properties

Besides being precipitated by polymers competing for water, gelatin is amphoteric, i.e. it has both positive and negative charges on the molecule (and no net charge at the isoionic point). Hence, at a pH where the basic side chains do not carry a charge, acid groups for example from gum arabic can react with the basic groups of gelatin to form an insoluble gelatin-arabate complex which can be precipitated around emulsified oil droplets, forming micro-ecapsulated oil. The microcapsules are hardened with formaldehyde or glutaraldehyde before harvesting and drying. In this application the pI of the gelatin is critical. This process has been used in the food industry for encapsulating flavours.

### 10.12.7 Milk - Food Stabilising Properties

Gelatin is used as a stabilizer particularly in yoghurt, where the addition of 0.3-0.5% acts to prevent syneresis thus allowing the production of stirred and fruit containing products. In this instance the gelatin reacts with the casein in the milk to reduce its tendency to separate water from the curd. Gelatin can also be used in cheese manufacture to improve yield and in the stabilisation of thickened cream.

### 10.12.8 Fruit Juice Clarifying Properties

In "fining" applications, gelatin reacts with polyphenols (tannins) and proteins in fruit juices forming a precipitate which settles leaving a supernatant which is stable to further cloud formation with storage time. In wine, usage levels are about 1 to 3 g/hL and excess usage, which would lead to protein instability, needs to be avoided. Traditionally, low Bloom strength gelatins are used but it has been shown that high Bloom strengths are equally effective. However, from the practical point of view, the use of low Bloom Strength gelatin is cheaper and makes it easier to mix the gelatin into the bulk of the cold juice before gelation can occur. In this regard, it has become common practice to treat cold grapes, during the initial crushing process, with gelatin that has been hydrolysed to the extent that it can no longer gel.

### 10.12.9 Texturising Properties

Gelatin is used in dried soups to provide the appropriate mouth feel (viscosity) to the final product.

### 10.12.10 Nutritional Properties

As stated earlier, gelatin is not a complete protein source because it is deficient in tryptophan and low in methionine content, however the digestibility is excellent and it is often used in feeding invalids and the high level of lysine (4%) is noteworthy. More controversially, studies have shown that the consumption of 7 to 10 g/day can significantly improve nail growth rate and strength and it also promotes hair growth. Gelatin has also been shown to benefit arthritis sufferers in a large proportion of cases.

### 10.12.11 Corrosive Properties

Although 304 stainless steel (s/s) can be used with milk, gelatin attacks 304 s/s and tubing can be perforated after a few months of continuous usage. With gelatin, it is essential to use 316 s/s and if heat exchanger plates are involved, the use of 316 s/s with the minimum specified molybdenum content of only 2% can be unacceptable.

## 10.13 FISH SKIN GELATIN

Fish skin gelatin is available commercially and can be produced for kosher use provided that the appropriate conditions are met (such as the use of fish having scales). Fish gelatin with normal gel strength has a normal hydroxyproline content and is made from fish from warmer waters and not necessarily from fresh water, although this is normally the case. Fish gelatin with low or no gel strength, has a low hydroxyproline content and is produced from cold water species which are sourced typically from the sea.

The low gel-strength gelatin has been used to emulsify vitamin A before spray drying to give another type of microencapsulated product using gelatin.

### 10.13.1 Testing Methods

The best published sources of gelatin testing methods are British Standard 757 of 1975 or Standard Methods for sampling and testing gelatin, published by the GMIA or the Pharmacopoeias. Many of the methods used in laboratories need to be modified to suit the peculiarities of gelatin.

### 10.13.2 Identification

Gelatin gives the normal positive trichloroacetic acid, biuret, ninhydrin tests for protein. The precipitate with 5% tannic acid is a particularly sensitive test for very dilute solutions of gelatin. In addition the thermally reversible gelation of a 6% solution in water between 10 and 60°C is unique for this protein.

### 10.13.3 Gel Strength

The most important attribute of gelatin is its gel strength and when determined by the standard method, is called the Bloom Strength. This is the force in grams required to press a 12.5 mm diameter plunger 4 mm into 112 g of a standard 62/3% w/v gelatin gel at 10°C. Several penetrometer type instruments have been adapted to determine Bloom Strength.

A frequent question is how to substitute gelatin of one Bloom Strength for a gelatin of another. As a guide one can say:

$$C \times B^{1/2} = k \ (24)$$

$$\text{or } C1(B1)^{1/2} \div (B2)^{1/2} = C2$$

Where C = concentration, B = Bloom strength, and k = constant, however, there are other considerations besides gel strength which can invalidate such a substitution calculation. For example, in a gummy formulation, the texture using 250 Bloom gelatin is far shorter than when 180 Bloom gelatin is used.

### 10.13.4 Viscosity

From the point of view of functionality, the solution viscosity of gelatin is probably the second most important parameter. The standard method calls for the viscosity of a 62/3 % solution at 60°C. Low viscosity (and a high gel strength) is required for poured confectionery, and high viscosity for film forming applications.

In viscosity calculations, usually $C \log V = k$, but the model is not as good as is the mathematical model for Bloom calculations.

### 10.13.5 Colour and Clarity

Solution colour and turbidity or clarity are attributes which may or may not be important depending on the application. Poor clarity markedly affects the ability to measure colour and at this stage there are no internationally accepted methods for determining these attributes, however, if clarity is good, then gelatin colour obeys Beer's Law.

### 10.13.6 pH

Solution pH (1%) is usually about pH 5 but can vary considerably. At this pH the viscosity of Type B gelatin is minimal and the gel strength is maximal, hence from the manufacturers point of view it is advantageous to manufacture gelatin at this pH. However, due to the strong buffering capacity of gelatin this pH may not be the most advantageous for the customer.

### 10.13.7 Moisture

The moisture content of gelatin may be as high as 16%, however, more normally it is about 10% to 13% because at 13.0% moisture content the glass transition temperature of gelatin is about 64°C which allows particle size reduction to be a simple operation. In addition, at 13% moisture content and 25°C gelatin is close to equilibrium with ambient air moisture contents of ca. 46% RH. At 6% to 8% moisture content gelatin is very hygroscopic and it becomes difficult to determine the physical attributes with accuracy.

Due to the variable granule size of gelatin, the rate of moisture loss at 105°C can be variable. Hence it is normal to add water to the gelatin powder before placing the sample in the drying oven. This means that the gelatin melts and water is lost from a uniform thin film of protein. It should be noted that metal dishes have to be used because, on drying, the film of gelatin shrinks and breaks containers of glass or ceramic.

Finally, the drying of gelatin to very low moisture contents results in cross-linking and loss of solubility. It is thus difficult to distinguish between free and bound water in gelatin.

### 10.13.8 Ash

The gelatin ash content is determined by pyrolysis at 550°C. Usually ash contents up to 2.5% can be accepted in food applications. However the nature of the ash can be important. For example, 2% CaSO4in gelatin can have excellent clarity in spite of the solubility product of the ash being exceeded (due to the crystal-habit modifying effect of gelatin), however on dilution of the gelatin in a confectionery formulation, the ash can precipitate. Furthermore, ammonia is often used as a pH modifier in gelatin preparation and salts like $NH_4Cl$ are not determinable by pyrolysis.

### 10.13.9 Sulphur Dioxide Content

Sulphur dioxide is used as a biocide and bleach in gelatin manufacture. The nationally permitted level of residual $SO_2$ in gelatin is variable and the methods for its determination can give a great variation in results. It is known that gelatin promotes oscillating redox reactions and the control of this contaminant is not easy. Hydrogen peroxide is often used to control the $SO_2$ content of gelatin and sometimes the permitted level of this contaminant is also specified. It is interesting to note that both $H_2O_2$ and $SO_2$ can be shown to coexist in gelatin.

### 10.13.10 Microbiological Properties

Gelatin is an excellent nutrient for most bacteria, hence the manufacturing processes have to carefully avoid contamination. Most countries have microbiological specifications for gelatin, but generally they are not very onerous. Total mesophyllic plate counts of 1000 are generally accepted with various countries limiting the presence of Coliforms, E. Coli, Salmonella, Clostridial spores, Staphylococci, and sometimes even Pseudomonades.

## 10.14 FOOD GELS

To build a model of protein gelation it helps to start with a model for proteins themselves. For most globular proteins we can start by ignoring the details of sequence and conformation and imagine most globular proteins as fairly dense particles a few nanometers in diameter with a hydrophobic core and a more hydrophilic surface; in effect colloidal particles rather than polymers. Depending on the protein these particles can be more or less water soluble but most gels require them to be able to form at least a stable suspension at a reasonable concentration (~10% protein).

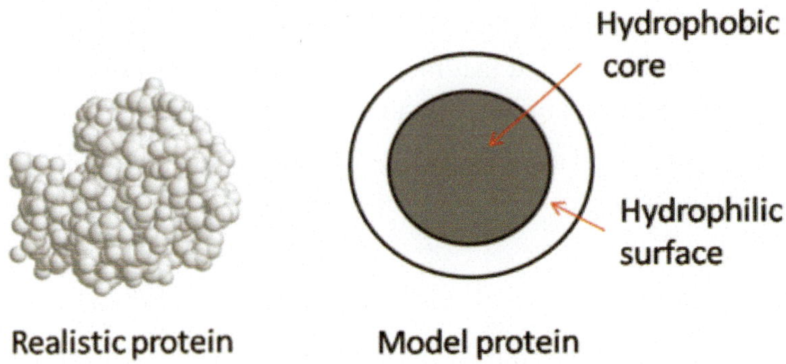


**Fig. 1 Most globular proteins**

To form a gel (i.e., trigger a sol-gel transition) the dispersion must be partly destabilized so the protein particles, rather than repelling one another, become somewhat "sticky" and aggregate. Different protein dispersions can be destabilized in different ways but common methods include denaturing the protein to properly expose the hydrophobic core (e.g., cooking egg), changing the pH to the isoelectric point to minimize inter-protein electrostatic repulsion (e.g., yogurt) or in some cases adding calcium to bind certain amino acid residues (e.g., forming a tofu gel) is useful. In other cases covalent bonds may form between proteins (e.g., forming gluten while kneading bread).

As destabilization proceeds, the aggregates become larger and larger and the suspension becomes more viscous. Eventually the aggregates will become so large they span the container giving an overall solid-like texture while trapping large amounts of water, i.e., a gel. If the associations between the proteins are too strong or too extensive then they may precipitate as a way of maximizing protein-protein interactions at the expense of the protein-water interactions needed in a gel.

The strength of the gel depends on the amount of material involved in the network and on the strength of the connections between proteins. Globular protein gels of this type typically need more polymer to form a gel than the polymer gels seen in polysaccharides and synthetic polymers (~10% vs 1-2%) and are often opaque as the protein aggregates can scatter light. Anything that disrupts the bonds between protein particles will weaken the gel.

# 11

# Fermented Foods Processing Flavor, Preservation, and HealtH

## 11.1 INTRODUCTION: AN OVERVIEW

Though fermentation is an ancient process used to prevent food from spoiling, it has recently exploded in popularity and has experienced rapid market growth due to reports that consuming fermented foods positively affects the gut microbiome and human health. The *New York Times*, *The Washington Post*, and the *Wall Street Journal* have all reported on the benefits of consuming fermented foods and their popularity on restaurant menus and on grocery store shelves. Fermentation is classified into six major types (Fig. 1) according to their end products. Among these some are typically employed in food preparation.

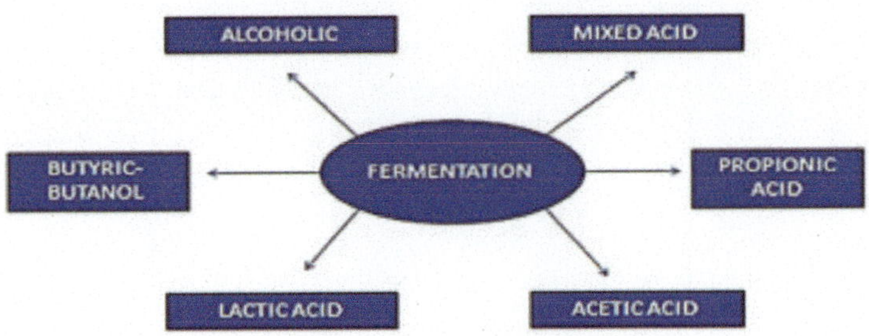


**Fig. 1 Types of fermentation**

The annual value of fermented foods in the U.S. market includes wine at >$60 billion, cheese at >$30 billion, beer at >$90 billion and total alcohol at >$250 billion, bread at >$30 billion, fermented vegetables at >$500 million, fermented meats at approximately $2 billion, yogurt at >$7 billion, kombucha at approximately $2 billion, and kefir at approximately $1 billion. Markets for all of these products are expected to continue to increase in value over the next decade, with extensive growth for yogurt, kefir, kombucha, and vegetables. The estimated global market for these specific products is between five and 15 times greater than what is reported for the U.S. market.

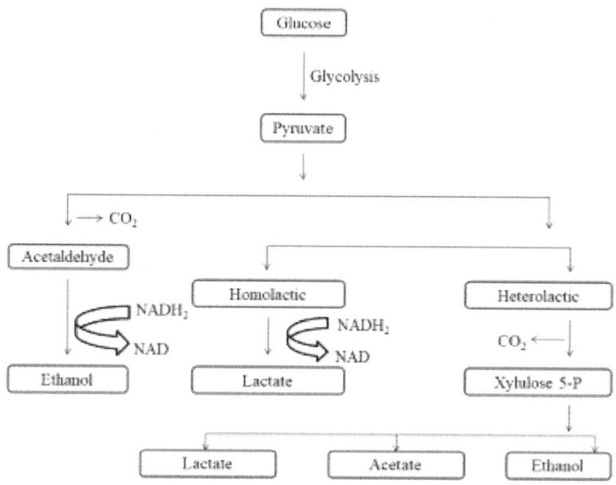


**Fig.2 Mechanism of fermentation**

## 11.2 FERMENTATION CHEMISTRY

Fermentation is a biochemical reaction that extracts energy from carbohydrates without using oxygen. Under anaerobic conditions, microbes break down sugar into alcohol, acid, and/or gas to produce adenosine triphosphate to stay alive, but the microbes also produce ethanol, carbon dioxide, and lactic acid, which increase product shelf life and impart desirable sensory properties.

For lactic acid bacteria, there are two parts to fermentation. First, glycolysis occurs in which the bacteria oxidize glucose and $2NAD^+$ reduces to 2NADH and produces two pyruvate molecules. In the second step of fermentation, pyruvate molecules are then reduced to lactic acid while NADH is oxidized to $NAD^+$, a process termed $NAD^+$ regeneration.

In the production of beer or wine, the yeast, such as *Saccharomyces cerevisiae,* will metabolize the sugar into pyruvate, in a process similar to lactic acid fermentation. However, with yeast, two carbon dioxide molecules are released from the pyruvate and acetaldehyde is formed. The acetaldehyde is then reduced to ethanol, while NADH is oxidized to $NAD^+$.

## 11.3 FERMENTATION FLAVOR

The desirable flavor of fermented foods is predominantly due to the acid, sugar, and volatile flavor compounds. As the bacteria and/or yeast consumes sugars, nonvolatile acids and volatile aroma compounds are formed. In sauerkraut, glucose and fructose are consumed by lactic acid bacteria (LAB), and the end product contains glucose, fructose, citric acid, malic acid, acetic acid, and lactic acid. The concentrations of these sugars, acids, and volatile aroma compounds contribute to the flavor (Satora et al. 2021). The volatile flavor compounds include esters, alcohols, carbonyl compounds, sulfur compounds, acids, terpenes, and nitriles, with the characteristic flavors of salty, sour, sulfur, pungent, and fermented (Satora et al. 2021). There will be different flavors associated with different fermented foods, but the unique tastes and aromas will be due to sugar, acid, and volatile flavor compound composition.

## 11.4 FERMENTATION FOR FOOD SUSTAINABILITY

Primary fermentation involves microorganisms, such as yeast and bacteria, enzymatically breaking down carbohydrates to produce carbon dioxide, organic acids, or alcohol. Fermentation contributes to food sustainability through preservation, safety, and production of food components. For example, wine has a pH of 3.0–4.0 due to acid content (tartaric, malic, citric, and lactic) and contains alcohol, which preserves the wine and makes it safe for consumption by inhibiting pathogen growth. An unopened bottle of wine will remain drinkable for a minimum of one year, but once opened will last less than a week. The wine will not be unsafe after one week, but will contain off-flavors.

Table-1: **List of Some Popular Fermented Foods and Microorganisms Used for Their Fermentation**

| Types of Fermented Foods | Name of Fermented Foods | Substrates of Fermented Foods | Types of Fermented Foods |
|---|---|---|---|
| Milk Products | Curd/Dadhi | Milk | *L. actobacillus bulgaricus,*<br>*L. cremoris, L. actococcus lactis* |
| | Butter Milk | Milk | *L. acidophilus, L. cremoris* |
| | Yoghurt | Milk | *L. thermophilus, L. bulgaricus* |
| | Camembert and Roquefort Cheese | Milk Casein | *Penicillium camemberti, P. roqueforti* |
| | Yukult | Milk | *L. actobacillus casei* |
| | Kefir | Milk | *L. caucasicus* |
| VegetableProducts | Sauerkraut Tempeh | Cabbage Soybean | *Leuconostoc mesenteroides*<br>*Aspergillus* sp. |
| Beverages | Sake | Rice | *Aspergillus oryzae* |
| | Wine | Grapes | *S. cerevisiae* |
| Food Additives | Vinegar | Alcohol | *Acetobacter* sp. And *Gluconobacter* sp. |

**TABLE .2 List of Some Traditional Fermented Foods**

| Sucuk | Meat | Turkey |
|---|---|---|
| Turşu, Salgam and Hardaliye | Fruits and Vegetables | |
| Tarhana | Cereals | |
| | | |
| Yoghurt, Torba Yoghurt, Kurut, Ayran, Kefir and Koumiss | | |
| | | |
| Milk | | |
| Yan-Jiang | Ginger | Taiwan |
| Jiang-Gua | Cucumber | |
| Jiang-Sun | Bamboo Shoot | |
| Yan-Tsai-Shin | Broccoli | |
| Kochujang | Soybean, Rice, Red Peeper | Korea |
| Kimchi | Vegetables | |
| Mabisi | Milk | Zambia |
| Munkoyo and Chibwantu | Cereals | |
| Adhirasam | Cereals | South India |
| Chakuli, Chitou and different kinds of Pitha | Cereals | Orissa, India |
| Kinema | Soybean | Sikkim and |
| | | |
| Gundruk | Leaves of Mustard | |
| Sinki | Fusiform Root of Radish | |
| Mesu | Bamboo Shoot | |
| Churpi | Milk | |
| Shel rod | Rice | |
| Jnards (Beer) | Rice | |

## 11.5 HEALTH BENEFITS

Consumption of fermented milk products with live cultures, such as yogurt and kefir, have been inversely correlated with type 2 diabetes, osteoporosis, impaired glucose metabolism, and decreased brain activity (Marco et al. 2017). Besides the phytochemicals, such as flavonoids and anthocyanins from the raw food materials, the live bacteria produce acids that help regulate glucose, blood pressure, diabetes, and cardiovascular diseases. The LAB in sauerkraut, including *Lactobacillus plantarum*, lessens inflammation and the presence of bad bacteria living within the gastrointestinal (GI) tract. Probiotic bacteria in fermented/ cultured foods provide health benefits, including reduced overall inflammation (both in and out of the GI tract), improvement of digestive disorders—like leaky gut syndrome and ulcerative colitis, improved immunity, and better nutrient absorption. Marco et al. (2020) reported that microbes in our diet, especially those from fermented foods, are linked to a reduced risk of cardiovascular disease, reduced risk of weight gain and type 2 diabetes, healthier metabolic profiles (blood lipids, glucose, pressure, and insulin resistance), and positively altered immune responses.

## 11.6 DAIRY FERMENTATION

Fermented dairy products are made from milk and include, but are not limited to, cheese, yogurt, and kefir. Yogurt and kefir contain live cultures similar to those present in probiotic supplements. The dairy food matrix buffers the pH of the stomach to protect the probiotic cultures so they are alive when they reach the small intestines. *Lactobacillus delbrueckii ssp. bulgaricus* and *ssp. lactis* are commonly used in freeze-dried yogurt starter cultures. These bacteria, along with *Bifidobacterium*, produce conjugated linoleic acids, vitamins, and bioactive peptides. Yogurt milk contains sucrose and/or glucose or a non-calorie sweetener and a stabilizer such as pectin, starch, or agar. The milk is generally standardized, deaerated, homogenized, and pasteurized prior to inoculation and fermentation. Kefir has traditionally been produced using kefir grains, which consist of a mixture of lactic acid bacteria, yeasts, and acetic acid bacteria in a polysaccharide and protein matrix. Most kefir produced today is made with a freeze-dried starter culture that contains a similar microbial composition to kefir grains. The diversity in bacteria and the presence of yeast provide a unique flavor different than yogurt, with a small percentage of ethanol. As we learn more about how probiotics and fermented foods containing live cultures beneficially impact the human microbiome and gut health, kefir and yogurt will likely continue to grow in the marketplace.

## 11.7 MEAT FERMENTATION

Examples of fermented meats include pepperoni, summer sausage, and salami. These products can last for months, if not years, due to culturing with *Micrococcus*, *Lactobacillus*, and/or *Leuconostoc*, which are used to bring the pH of the meat to approximately 5.0 over time so that the pH is at the approximate isoelectric of the salt-soluble proteins, myosin and actin. This causes moisture loss such that the water activity and salt content make the product shelf stable. This is a relatively short process of 2–3 days for semidry sausages, like summer sausage. But for dry sausages, like pepperoni and hard salami, it can take 30 days or longer, and more than 90 days for genoa, in which the green room, or fermentation room, has specific temperatures, relative humidity, and air flow to dry the meat and optimize quality.

## 11.8 VEGETABLE FERMENTATION

Two fermentation products produced from cabbage include sauerkraut and kimchi. Sauerkraut can be made through natural fermentation or through the addition of a starter culture, while kimchi is made through spontaneous fermentation from vegetables, including cabbage, chives, radishes, and leeks. Kimchi has become a hugely popular consumer good. According to Mike Maney, the director of technical services at GLK Foods, the largest sauerkraut producer in the world, "All of our sauerkraut is naturally fermented. We cut in excess of 120 tons of cabbage per hour and utilize high-speed color/vision sorters and lasers to sort out defects in the sliced cabbage. We automatically add salt in-line by weight to ensure product quality. We are confident that there will be continued growth in all fermented foods, especially sauerkraut. Every day there seems to be new findings regarding the benefits of fermented foods with respect to gut health, reducing inflammation in the body, and antioxidant properties."

## 11.9 LOW-INTERVENTION WINE

Low-intervention, or natural, wine is produced from hand-picked grapes that are fermented using yeast native to the environment, with small quantities of sulfites and no other added ingredients. Conventional wine is commonly made from machine-harvested grapes with the following added ingredients: 1) yeast to control the fermentation process and flavor, 2) acid to increase acidity to help the aging process, and 3) sulfites at bottling (up to 350 ppm) to stop fermentation and preserve flavor. Gustavo A. Sotelo is the winemaker at Scribe, a vineyard that produces low-intervention wines. According to Sotelo, "The global wine movement for low-intervention wines has been on the rise for the last decade. All our wine fermentation occurs naturally, using indigenous yeast coming in from the vineyard. Our sparkling, white wine, and rosé wines are fermented cold to retain purity and freshness in the wines. Our skin-fermented chardonnay and our red wine fermentations occur much warmer and in open-top tanks to extract color, flavor, and aroma from the skins."

## 11.10 THE FUTURE OF FERMENTED FOODS

Future trends in fermentation will include a greater focus on natural fermentation and increased growth in the craft beer, kombucha, vegetable fermentation, yogurt, and kefir markets. There will also be increased growth in fermented food markets as researchers learn more about the relationship between the composition of different fermented foods, the human gut microbiome, and human health. Whether it is for health reasons, flavor, preservation, environmental concerns, or a combination of these reasons, fermented foods will continue to thrive and increase in the United States and in global markets. In addition, with the increased availability of kombucha, sauerkraut, and craft beer kits, there will also be an increase in the home production of fermented foods.

# BIBLIOGRAPHY


- Admassie, M. (2018). A Review on Food Fermentation and the Biotechnology of Lactic Acid Bacteria. World Journal of Food Science and Technology, 2(1), 19. https://doi.org/10.11648/j.wjfst.20180201.13.
- Ajeet Kumar Srivastava, PushpaAgrawal, Abdul Rahiman,Delignification of Rice Husk and Production of Bioethanol. 2014.
- Ajeet Kumar Srivastava, PushpaAgrawal, Abdul Rahiman,Delignification of Rice Husk and Production of Bioethanol.
- Al-Sheraji, S.H., et al., Prebiotics as functional foods: a review. Journal of Functional Foods, 2013. 5(4):1542-1553.
- Bell, V., J. Ferrão, and T. Fernandes, Nutritional Guidelines and Fermented Food Frameworks. Foods, 2017. 6(8):65.
- ChandrakantBelwal, Divyesh Patel, KamleshChauhan, YogendrasinhParmar, Ajay Singh Rawat, et. al.Isolation, Identification and Characterization of Unknown Impurity in Fermentation Based Active Pharmaceutical Ingredient Lovastatin. 2014;Pharmaceutical Analysis.
- Chen, M., et al., Dairy consumption and risk of type 2 diabetes: 3 cohorts of US adults and an updated meta-analysis. BMC medicine, 2014. 12(1):215.
- Cho, Y.A., Effect of Probiotics on Blood Lipis Concentrations: A meta-analysis of randomized controlled trials. Medicine, 2015. 94(43):e1714
- Ciani, M., Comitini, F., & Mannazzu, I. (2018). Fermentation. Encyclopedia of Ecology, June, 310–321. https://doi.org/10.1016/B978-0-12-409548-9.00693-X.
- Diop MB, Destain J, Alvarez VB, Konte MA and Thonart P, Use of Nisin-Producing Starter Cultures of Lactococcuslactis subsp. lactis on Cereal Based-Matrix to Optimize Preservative Factors over Fish Fermentation at 30°C Typical to Senegal.J Food Process Technol 2015; 6:432.

- Eric L Huang, Mark G Lefsrud,Fermentation Monitoring of a Co-Culture Process with Saccharomyces cerevisiae and Scheffersomycesstipitis Using Shotgun Proteomics.J Bioprocess Biotech 2014; 4:144
- EsraAydemir, SelamiDemirci, Ayeguumll Doan, Ali OumlzhanAytekin and FikrettinSahin, Genetic Modifications of Saccharomyces cerevisiae for Ethanol Production from Starch Fermentation: A Review.J Bioprocess Biotech 2014; 4:180.
- Eussen, S.J., et al., Consumption of dairy foods in relation to impaired glucose metabolism and type 2 diabetes mellitus: the Maastricht Study. British Journal of Nutrition, 2016. 115(8):1453-1461.
- F. O. Faithpraise, B. Usibe,,J.Idung, C. R. Chatwin, et. al. Pesticide Free Control of Mosquitoes via Toxorhynchites predators and Fermentation Traps. 2014;IJIRSET
- Fruton, Joseph S. Proteins, Enzymes, Genes: The Interplay of Chemistry and Biology. Yale University Press: New Haven, 1999. ISBN 0-300-07608-8.
- Ghosh, B., Bhattacharya, D., & Mukhopadhyay, M. (2018). Use of Fermentation Technology for Value-Added Industrial Research. Principles and Applications of Fermentation Technology, August, 141–161. https://doi.org/10.1002/9781119460381.ch8.
- Gibson, G.R., et al., Expert consensus document: The International Scientific Association for Probiotics and Prebiotics (ISAPP) consensus statement on the definition and scope of prebiotics. Nature Reviews Gastroenterology and Hepatology, 2017. 14(8):491.
- GunjanGautam, Vishwas Mishra, PayalVerma, Ajay Kumar Pandey and SangeetaNegi, A Cost Effective Strategy for Production of Bio-surfactant from Locally Isolated Penicilliumchrysogenum SNP5 and Its Applications.J Bioprocess Biotech 2014; 4:177
- Hill, C., et al., Expert consensus document: The International Scientific Association for Probiotics and Prebiotics consensus statement on the scope and appropriate use of the term probiotic. Nature Reviews Gastroenterology and Hepatology, 2014. 11(8):506.
- Hind, H. L., & Day, F. E. (1930). Fermentation Industries. In Journal of the Institute of Brewing (Vol. 36, Issue 6, pp. 1–29). https://doi.org/10.1002/j.2050-0416.1930.tb05286.x.
- Hooi Ling Ho and JeeHsiungPhang,Bioprocessing of Agro-Residual Wastes for Optimisation of Xylanase Production by Aspergillusbrasiliensis in Shake Flask Culture and Its Scaling up Elucidation in Stirred Tank Bioreactor.J BiodiversBioprosDev 2015; 2:148.

- Hooi Ling Ho, Effects of Medium Formulation and Culture Conditions on Microbial Xylanase Production Using Agricultural Extracts in Submerged Fermentation (SmF) and Solid State Fermentation (SsF): A Review.J BiodiversBioprosDev 2014; 1:130
- Hooi Ling Ho, Xylanase Production by Bacillus subtilis Using Carbon Source of Inexpensive Agricultural Wastes in Two Different Approaches of Submerged Fermentation (SmF) and Solid State Fermentation (SsF).J Food Process Technol 2015; 6:437.
- Hotel, A.C.P. and A. Cordoba, Health and nutritional properties of probiotics in food including powder milk with live lactic acid bacteria. Prevention, 2001. 5(1):1-34.
- Ibrahim, AliyuDabai Aisha Ibrahim Saulawa, AlhassanSani Danladi Mahuta Sahabi Saadatu Aliyu Shinkafi Adamu Aliyu Aliero and Gambo, Auwal, bioutilization of adansonia digitata fruit pulp by bacillus species for amylase production.
- Iwasa, M., et al., Fermented milk improves glucose metabolism in exercise-induced muscle damage in young healthy men. Nutrition journal, 2013. 12(1):83.
- JahirAlam Khan and Sachin Kumar Yadav, production of alpha amylases by aspergillus niger using cheaper substrates employing solid state fermentation. 2011.
- JahirAlam Khan and Sumit Kumar Singh, production of cellulase using cheap substrates by solid state fermentation. 2011.
- KesavapillaiBalakrishnan, Aiswarya Nair and Rahul Kumar, Screening of Microbial Isolates for the Fermentative Production of L-Asparaginase in Submerged Fermentation. Journal of Pharmacy and Pharmaceutical Sciences, 2013.
- Kohler, Robert. From Medical Chemistry to Biochemistry: The Making of a Biomedical Discipline. Cambridge University Press, 1982.
- Landine, R., De Garie, C., & Cocci, A. (1997). Fermentation process. Biotechnology Advances, 15(3–4), 702.
- Marco, M.L., et al., Health benefits of fermented foods: microbiota and beyond. Current opinion in biotechnology, 2017. 44:94-102.
- MarekNawalany and MalgorzataLoga, Fermentation Tube Test Statistics for Indirect Water Sampling.Hydrol Current Res 2014; 5:165
- Maria Papagianni, Characterization of Fungal Morphology using Digital Image Analysis Techniques.J MicrobBiochemTechnol 2014; 6:189- 194

- Martínez-Espinosa, R. M. (2020). Introductory Chapter: A Brief Overview on Fermentation and Challenges for the Next Future. New Advances on Fermentation Processes.
- Mervat MA ElGendy and Ahmed MA ElBondkly, Optimization of Solid State Fermentation and Leaching Process Parameters for Improvement Xylanase Production by Endophytic Streptomyces sp. ESRAA-301097.J Microb Biochem Technol2014; 6:154-166
- Microbiology, F. (2016). Basic Principles of Food Fermentation. Food Microbiology: Principles into Practice, 228–252.
- NavpreetKaurWalia, KamaljeetKaurSekhon, Swaranjit Singh Cameotra, Dharam Paul Chaudhary, PallaviSrivastava and Anil Dutta, et. al. Optimization of Fermentation Parameters for Bioconversion of Corn to Ethanol Using Response Surface Methodology.J Pet Environ Biotechnol 201;, 5:178
- Peddapalli Siva Rasagnya And MeenaVangalapati, Studies on Optimization of Process Parameters for Nattokinase Production by Bacillus subtilis NCIM 2724 and Purification by Liquid-Liquid Extraction 2013.
- Principles and Applications of Fermentation Technology. (2018). In Principles and Applications of Fermentation Technology.
- S.A Thakur,S.N.Nemade, Sharanappa A., Solid State Fermentation of Overheated Soybean Meal (Waste) For Production of Protease Using AspergillusOryzae. 2015.
- Şanlier, N., B.B. GÖkcen, and A.C. Sezgin, Health benefits of fermented foods. Critical reviews in food science and nutrition, 2017:1-22.
- Sharma, R., Garg, P., Kumar, P., Bhatia, S. K., & Kulshrestha, S. (2020). Microbial fermentation and its role in quality improvement of fermented foods. Fermentation, 6(4), 1–20.
- Soedamah-Muthu, S.S., et al., Consumption of dairy products and associations with incident diabetes, CHD and mortality in the Whitehall II study. British Journal of Nutrition, 2013. 109(4):718-726.
- SumaryatiSyukur, BenrwardBisping, ZozyAneloiNoli and EndangPurwati, Antimicrobial Properties and Lactase Activities from Selected Probiotic Lactobacillus brevis Associated With Green Cacao Fermentation in West Sumatra, Indonesia. J Prob Health 2013; 1: 113
- Tania Pencheva, Maria Angelova, Purposeful Model Parameters Genesis in Multi-population Genetic Algorithm.2012; Glob J Tech Opt
- Wei Jiang, Zhao Li, Hongqiang Li and Jian Xu, Effect of Different Sweet Sorghum Storage Conditions on Ethanol Production.BiochemPhysiol 2014;3: 142.